# Introduction to GIS Programming and Fundamentals with Python and ArcGIS®

# Introduction to GIS Programming and Fundamentals with Python and ArcGIS®

Chaowei Yang

With the collaboration of

**Manzhu Yu**

**Qunying Huang**

**Zhenlong Li**

**Min Sun**

**Kai Liu**

**Yongyao Jiang**

**Jizhe Xia**

**Fei Hu**

CRC Press
Taylor & Francis Group
Boca Raton London New York

CRC Press is an imprint of the
Taylor & Francis Group, an **informa** business

CRC Press
Taylor & Francis Group
6000 Broken Sound Parkway NW, Suite 300
Boca Raton, FL 33487-2742

© 2017 by Taylor & Francis Group, LLC
CRC Press is an imprint of Taylor & Francis Group, an Informa business

No claim to original U.S. Government works

Printed on acid-free paper

International Standard Book Number-13: 978-1-4665-1008-1 (Hardback)

### Library of Congress Cataloging-in-Publication Data

Names: Yang, Chaowei, author.
Title: Introduction to GIS programming and fundamentals with Python and ArcGIS / Chaowei Yang.
Description: Boca Raton, FL : Taylor & Francis, 2017.
Identifiers: LCCN 2016047660| ISBN 9781466510081 (hardback : alk. paper) | ISBN 9781466510098 (ebook)
Subjects: LCSH: Geographic information systems--Design. | Python (Computer program language) | ArcGIS
Classification: LCC G70.212 .Y36 2017 | DDC 910.285/53--dc23
LC record available at https://lccn.loc.gov/2016047660

**Visit the Taylor & Francis Web site at**
**http://www.taylorandfrancis.com**

**and the CRC Press Web site at**
**http://www.crcpress.com**

*For Chaowei Yang's parents, Chaoqing Yang and Mingju Tang,*

*for continually instilling curiosity and an exploring spirit*

# Contents

## Section III   Advanced GIS Algorithms and Their Programming in ArcGIS

# Section IV   Advanced Topics

# *Preface*

## Why Another GIS Programming Text?

Geographical information system (GIS) has become a popular tool underpinning many aspects of our daily life from routing for transportation to finding a restaurant to responding to emergencies. Convenient GIS tools are developed with different levels of programming from scripting, using python for ArcGIS, to crafting new suites of tools from scratch. How much programming is needed for projects largely depends on the GIS software, types of applications, and knowledge structure and background of the application designer and developer. For example, simple scripting integrates online mapping applications using Google maps. Customized spatial analyses applications are routinely using ArcGIS with minimum programming. Many develop an application leveraging open-source software for managing big data, modeling complex phenomena, or responding to concurrent users for popular online systems. The best design and development of such applications require designers and developers to have a thorough understanding of GIS principles as well as the skill to choose between commercial and open-source software options. For most GIS professionals, this is a challenge because most are either GIS tool end users or information technology (IT) professionals with a limited understanding of GIS.

To fill this gap, over the last decade, Chaowei Yang launched an introductory GIS programming course that was well received. Enrollment continues to rise and students report positive feedback once they are in the workplace and use knowledge developed from the class. To benefit a broader spectrum of students and professionals looking for training materials to build GIS programming capabilities, this book is written to integrate and refine the authors' knowledge accumulated through courses and associated research projects.

The audience for this book is both IT professionals to learn the GIS principles and GIS users to develop programming skills. On the one hand, this book provides a bridge for GIS students and professionals to learn and practice programming. On the other hand, it also helps IT professionals with programming experience to acquire the fundamentals of GIS to better hone their programming skills for GIS development.

Rather than try to compete with the current GIS programming literature, the authors endeavor to interpret GIS from a different angle by integrating GIS algorithms and programming. As a result, this book provides a practical knowledge that includes fundamental GIS principles, basic programming skills, open-source GIS development, ArcGIS development, and advanced topics. Structured for developing GIS functions, applications, and systems, this book is expected to help GIS/IT students and professionals to become more competitive in the job market of GIS and IT industry with needed programming skills.

## What Is Included in the Text?

This book has four sections. Section I (Chapters 1 and 2) is an overview of GIS programming and introduces computer and programming from a practical perspective. Python (integral programming language for ArcGIS) programming is extensively presented in Section II (Chapters 3 through 8) in the context of designing and developing a Mini-GIS using hands-on experience following explanations of fundamental concepts of GIS. Section III (Chapters 9 through 12) focuses on advanced GIS algorithms and information on how to invoke them for programming in ArcGIS. Advanced topics and performance optimization are introduced in Section IV (Chapters 13 and 14) using the Mini-GIS developed.

Chapter 1 introduces computer, computer programming, and GIS. In addition, the Unified Markup Language (UML) is discussed for capturing GIS models implemented through simple Python programming. Chapter 2 introduces object-oriented programming and characteristics with examples of basic GIS vector data types of Point, Polyline, and Polygon.

Chapter 3 introduces Python syntax, operators, statements, miscellaneous features of functions, and Python support for object-oriented programming. Using GIS examples, Chapter 4 introduces Python language control structures, file input/output, and exception handling. Chapter 5 presents programming thinking using the visualization of vector data as an example of the workflow of this critical process in programming. Chapter 6 introduces the Python integrated programming environment (IDE), modules, package management, and the Mini-GIS package. Chapter 7 discusses shapefile formats and steps on how to handle shapefiles within the Mini-GIS. Chapter 8 introduces vector data processing algorithms and includes line intersection, centroid, area, length, and point in polygon. This presentation includes how Mini-GIS/ArcGIS supports these algorithms.

Chapter 9 bridges Sections II and III by introducing ArcGIS programming in Python using ArcPy, ArcGIS programming environment, automating tools, accessing data, describing objects, and fixing errors. Chapter 10 introduces raster data algorithms, including raster data format, storage, and compression with hands-on experience using ArcGIS. Chapter 11 addresses network data algorithms for representing networks and calculating the shortest path in principles and using ArcGIS. Chapter 12 explores surface or

3D data representation of 3D data, converting data formats and 3D analyses for elevation, slope, aspect, and flow direction with examples in ArcGIS programming.

Chapter 13 introduces performance-improving techniques and includes storage access and management, parallel processing and multithreading, spatial index, and other techniques for accelerating GIS as demonstrated in Mini-GIS. Advanced topics, including GIS algorithms and modeling, spatial data structure, distributed GIS, spatiotemporal thinking, and computing, are presented in Chapter 14.

## Hands-On Experience

As a practical text for developing programming skills, this book makes every effort to ensure the content is as functional as possible. For every introduced GIS fundamental principle, algorithm and element, an example is explored as a hands-on experience using Mini-GIS and/or ArcGIS with Python. This learning workflow helps build a thorough understanding of the fundamentals and naturally maps to the fundamentals and programming skills.

For system and open-source development, a step-by-step development of a python-based Mini-GIS is presented. For application development, ArcGIS is adopted for illustration.

The Mini-GIS is an open-source software developed for this text and can be adopted for building other GIS applications. ArcGIS, a commercial product from ESRI, is used to experience state-of-the-art commercial software. For learning purpose, ArcGIS is available for free from ESRI.

## Online Materials

This book comes with the following online materials:

- Instructional slides for instructors using this text for classroom education and professionals to assist in learning GIS programming.
- Python codes for class exercises and hands-on experiences and structured and labeled by chapter to code the chapter's sequence.
- Mini-GIS as an open-source package for learning the GIS fundamentals and for exemplifying GIS principles and algorithms.
- Answers to problems for instructors to check their solutions.

## The Audience for and How to Use This Text

This text serves two functions: a text for systematic building GIS programming skills and a reference for identifying a python solution for specific GIS algorithms or function from scratch and/or ArcGIS. The text is intended to assist four categories of readers:

- Professors teaching GIS programming or GIS students learning with a specific focus on hands-on experience in classroom settings.
- Programmers wanting to learn GIS programming by scanning through Section I and Chapters 3 and 4, followed by a step-by-step study of the remaining chapters.
- GIS system designers most interested in algorithm descriptions, algorithms implementation from both scratch and ArcGIS to assemble a practical knowledge about GIS programing to aid in GIS choice for future development.
- IT professionals with a curiosity of GIS for GIS principles but skipping the programming exercises.

The intent of the authors for such a broad audience is based on the desire to cultivate a competitive professional workforce in GIS development, enhance the literature of GIS, and serve as a practical introduction to GIS research.

## How Did We Develop This Text?

The text material was first developed by Professor Chaowei Yang in 2004 and offered annually in a classroom setting during the past decade. During that time span, many students developed and advanced their programming skills. Some became professors and lecturers in colleges and were invited to write specific book chapters. Keeping the audience in mind, several professors who teach GIS programming in different cultural backgrounds and university settings were invited to review the book chapters.

The following is the book development workflow:

- Using his course materials, Professor Yang structured this book with Irma Shagla's help, and the text's structure was contracted to be published as a book. Assistant Professor Qunying Huang, University of Wisconsin, Madison, explored using the earlier versions of the text's materials. Assistant Professors Huang and Zhenlong Li, University of South Carolina, developed Section II of the text in collaboration with Professor Yang.

- Dr. Min Sun, Ms. Manzhu Yu, Mr. Yongyao Jiang, and Mr. Jizhe Xia developed Section III in collaboration with Professor Yang.
- Professor Yang edited and revised all chapters to assure a common structure and composition.
- Ms. Manzhu Yu and Professor Yang edited the course slides.
- Assistant Professor Li, Mr. Kai Liu, Mrs. Joseph George, and Ms. Zifu Wang edited Mini-GIS as the software for the text.
- After the above text and course materials were completed, four professors and two developers were invited to review the text's content.
- The assembled materials for the text were finally reviewed by several professionals, including Ms. Alena Deveau, Mr. Rob Culbertson, and Professor George Taylor.
- The text was formatted by Ms. Minni Song.
- Ms. Manzhu Yu and Professor Yang completed a final review of the chapters, slides, codes, data, and all relevant materials.

# Acknowledgments

This text is a long-term project evolving from the course "Introduction to GIS Programming" developed and refined over the past decade at George Mason University. Many students and professors provided constructive suggestions about what to include, how best to communicate and challenge the students, and who should be considered as audience of the text.

The outcome reflects Professor Yang's programming career since his undergraduate theses at China's Northeastern University under the mentoring of Professor Jinxing Wang. Professor Yang was further mentored in programming in the GIS domain by Professors Qi Li and Jicheng Chen. His academic mentors in the United States, Professors David Wong and Menas Kafatos, provided support over many decades, giving him the chance to teach the course that eventually led to this text.

Professor Yang thanks the brilliant and enthusiastic students in his classes at George Mason University. Their questions and critiques honed his teaching skills, improved the content, and prompted this effort of developing a text.

Professor Yang thanks his beloved wife, Yan Xiang, and children—Andrew, Christopher, and Hannah—for accommodating him when stealing valuable family time to complete the text.

Ms. Manzhu Yu extends her gratitude to the many colleagues who provided support, and read, wrote, commented, and assisted in the editing, proofreading, and formatting of the text.

Assistant Professor Huang thanks her wonderful husband, Yunfeng Jiang, and lovely daughter, Alica Jiang.

Dr. Min Sun thanks her PhD supervisor, Professor David Wong, for educating her. She also thanks David Wynne, her supervisor in ESRI where she worked as an intern, and her other coworkers who collectively helped her gain a more complete understanding of programming with ESRI products. Last but not least, she thanks her parents and lovely dog who accompanied her when she was writing the text.

Yongyao Jiang thank his wife Rui Dong, his daughter Laura, and his parents Lixia Yao and Yanqing Jiang.

# *Editor*

**Chaowei Yang** is a professor of geographic information science at George Mason University (GMU). His research interest is on utilizing spatiotemporal principles to optimize computing infrastructure to support science discoveries. He founded the Center for Intelligent Spatial Computing and the NSF Spatiotemporal Innovation Center. He served as PI or Co-I for projects totaling more than $40 M and funded by more than 15 agencies, organizations, and companies. He has published 150+ articles and developed a number of GIS courses and training programs. He has advised 20+ postdoctoral and PhD students who serve as professors and scientists in highly acclaimed U.S. and Chinese institutions. He received many national and international awards, such as the U.S. Presidential Environment Protection Stewardship Award in 2009. All his achievements are based on his practical knowledge of GIS and geospatial information systems. This book is a collection of such practical knowledge on how to develop GIS tools from a programming perspective. The content was offered in his programming and GIS algorithm classes during the past 10+ years (2004–2016) and has been adopted by his students and colleagues serving as professors at many universities in the United States and internationally.

# Contributors

**Fei Hu** is a PhD candidate at the NSF Spatiotemporal Innovation Center, George Mason University. He is interested in utilizing high-performance cloud computing technologies to manage and mine big spatiotemporal data. More specifically, he has optimized the distributed storage system (e.g., HDFS) and parallel computing framework (e.g., Spark, MapReduce) to efficiently manage, query, and analyze big multiple-dimensional array-based datasets (e.g., climate data and remote sensing data). He aims to provide scientists with on-demand data analytical capabilities to relieve them from time-consuming computational tasks.

**Qunying Huang** is an assistant professor in the Department of Geography at the University of Wisconsin, Madison. Her fields of expertise include geographic information science (GIScience), cyber infrastructure, spatiotemporal big data mining, and large-scale environmental modeling and simulation. She is very interested in applying different computing models, such as cluster, grid, GPU, citizen computing, and especially cloud computing, to address contemporary big data and computing challenges in the GIScience. Most recently, she is leveraging and mining social media data for various applications, such as emergency response, disaster mitigation, and human mobility. She has published more than 50 scientific articles and edited two books.

**Yongyao Jiang** is a PhD candidate in Earth systems and geoinformation sciences at the NSF Spatiotemporal Innovation Center, George Mason University. He earned an MS (2014) in GIScience at Clark University and a BE (2012) in remote sensing at Wuhan University. His research focuses on data discovery, data mining, semantics, and cloud computing. Jiang has received the NSF EarthCube Visiting Graduate Student Early-Career Scientist Award (2016), the Microsoft Azure for Research Award (2015), and first prize in the Robert Raskin CyberGIS Student Competition (2015). He serves as the technical lead for MUDROD, a semantic discovery and search engine project funded by NASA's AIST Program.

**Zhenlong Li** is an assistant professor in the Department of Geography at the University of South Carolina. Dr. Li's research focuses on spatial high-performance computing, big data processing/mining, and geospatial cyberinfrastructure in the area of data and computational intensive GISciences. Dr. Li's research aims to optimize spatial computing infrastructure by integrating cutting-edge computing technologies and spatial principles to support domain applications such as climate change and hazard management.

**Kai Liu** is a graduate student in the Department of Geography and GeoInformation Sciences (GGS) in the College of Science at George Mason University. Previously, he was a visiting scholar at the Center of Intelligent Spatial Computing for Water/Energy Science (CISC) and worked for 4 years at Heilongjiang Bureau of Surveying and mapping in China. He earned a BA in geographic information science at Wuhan University, China. His research focuses on geospatial semantics, geospatial metadata management, spatio-temporal cloud computing, and citizen science.

**Min Sun** is a research assistant professor in the Department of Geography and Geoinformation Science at George Mason University. Her research interests include measuring attribute uncertainty in spatial data, developing visual analytics to support data exploration, WebGIS, and cloud computing. She is an expert in ArcGIS programming and also serves as the assistant director for the U.S. NSF Spatiotemporal Innovation Center.

**Jizhe Xia** is a research assistant professor at George Mason University. He earned a PhD in Earth systems and geoinformation sciences at the George Mason University in the spring of 2015. Dr. Xia's research interests are spatiotemporal computing, cloud computing, and their applications in geographical sciences. He proposed a variety of methods to utilize spatiotemporal patterns to optimize big data access, service quality (QoS) evaluation, and cloud computing application.

**Manzhu Yu** is a PhD candidate in the Department of Geography and Geoinformation Science, George Mason University. Her research interests include spatiotemporal methodology, pattern detection, and spatiotemporal applications on natural disasters. She received a Presidential Scholarship from 2012 to 2015. She has published approximately 10 articles in renowned journals, such as *PLoS ONE* and *IJGIS*, and contributed as a major author in several book chapters.

# Section I

# Overview

# 1

## Introduction

This chapter introduces the basic concepts of computer, hardware, software, and programming, and sets up the context for GIS programming.

## 1.1 Computer Hardware and Software

A *computer* is a device that has the capability to conduct different types of automated tasks based on specific instructions predefined by or through interactions with end users. For example, clicking on the ArcGIS icon will execute ArcGIS software. We can select a destination and starting point to trigger a routing analysis to identify a driving route using Google Maps. Computers are some of the fastest-evolving technologies as reflected by the processing capability of small calculators to supercomputers. The size of the devices has reduced from computers occupying a building to mobile devices in pockets (Figure 1.1). The user interactions range from typing punched cards (early computers) to human–computer interaction, such as speaking to invoke an action or task.

There are two important components of a computer (Hwang and Faye 1984): (1) the physical device that can conduct automated processing, and (2) instruction packages that can be configured to provide specific functionality, such as word processing or geographic information processing. The first component of a computer, the *hardware*, is touchable as physical machines. The second component, the *software*, may be purchased with the hardware in the form of an operating system, or installed by downloading online. Computer hardware can be configured or programmed to perform different tasks; thus, a computer may also be called a general-purpose device. The software varies greatly, whether it is providing document-processing capability, financial management, tax return processing, or scientific simulations such as climate change or the spread of disease. Depending on the type of software, it is either procured publicly (freeware) or proprietary (requiring purchase and licensing). Depending on the usage, software can be categorized as system software, application software, or embedded software (Figure 1.2). System software refers to the basic software that must be installed for a computer to operate. Windows and Linux are examples of operating system (OS) software, an essential component of a computer. Application software supports

(a)                                             (b)

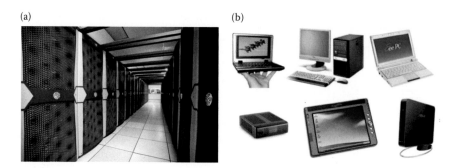

**FIGURE 1.1**
(a) NASA supercomputer. (From NASA supercomputer at http://www.nas.nasa.gov/hecc/resources/pleiades.html.) (b) Other computers: personal computer (PC), laptop, pad. (From different computers at http://www.computerdoc.com.au/what-are-the-different-types-of-computers.)

specific groups of tasks, such as Microsoft Word for document processing and Microsoft Outlook for emails. Embedded software is a type of firmware that is burned onto hardware and becomes part of that hardware. Embedded software exists longer on a computer than any other software. The firmware will always come with the hardware when you purchase a computer, so the firmware will not have to be changed as frequently, especially when updating a web browser or Turbo Tax Return routinely.

*Geographic information system* (GIS) is one type of application software that deals primarily with geographic information (Longley et al. 2001). The global positioning system (GPS, Misra and Enge 2006) is used for locating geographic places, and can be installed in both cars and smart phones for routing. GIS software includes two categories: professional GIS and lightweight GIS. Professional GIS software, such as ArcGIS, provides the most

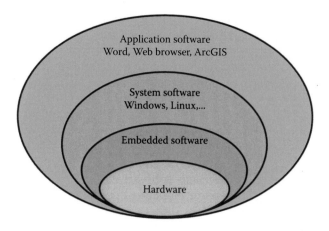

**FIGURE 1.2**
Different types of software.

complete set of GIS functionalities for professionals in the GIS domain. Less intense, but popular, GIS software used to view the geographic environment are the online mapping application, such as Google Maps and Google Earth.

## 1.2 GIS and Programming

GIS originates from several domains and refers to the system designed to capture, observe, collect, store, and manage geographic data, and to provide tools for spatial analyses and visualization (Longley et al. 2001). GIS can help obtain geographic data to be used for decision making, such as choosing routes for emergency response. GIS is known to have started from the Canadian natural resource inventory computer program led by Roger Tomlinson in the 1960s. GIS is becoming increasingly popular on mobile devices as a means of analyzing information and patterns related to traffic and weather.

Coined by Mike Goodchild, the term "GIS" can also refer to the field of geographic information science or *GIScience*—the study of the scientifically applied GIS principles and technologies (Goodchild 1992). According to GIS scientists, *GIScience* involves remote sensing, global navigation satellite systems, and GIS. Additionally, in various domains, *GeoInformatics* may be applied to remote sensing, global navigation satellite system, and GIS information. These topics, however, will not be explored in this book.

GIS is the system comprising hardware (computer, mobile devices, GPS), software (ArcGIS or online mapping), and data (geographic information) that can be utilized to accomplish a set of functionalities for a group of users. All three components must be utilized for GIS to work effectively. A significant difference between GIS and other software applications is its ability to manage and manipulate the large volume and complexity of geographic data, which comprises embedded spatiotemporal and attribute information. The complex character of GIS data demands a specific suite of software to extract information for decision making. Mature software packages are publicly available, including the most up-to-date set of ArcGIS software and the latest edition of Google Maps web mapping software.

The process of developing software is called *programming*. Programming instructs the computer to accomplish a task based on the orders. There are many different types of programming levels (Mitchell 1996). The lowest level to program are based on the specific hardware instructions supported by the central processing units (CPU), and used by smart-instrument developers. Because CPU instructions are processed as a sequence of 0s and 1s, assembling language is developed to assist developers to remember those instructions. Both languages are considered low level and are specific to the hardware. Advanced languages have been developed to facilitate human understanding, but are still restricted by the hardware instructions. For example, C programming language is commonly used to develop software (Kernighan and Ritchie 2006). To make the programming organization more similar to how we view the world, C++ was proposed to support object-oriented programming based on C (Stroustrup 1995). Since then, many different programming languages have been developed and are used in GIS programming. For instance, Java is a language for cross-platform application development proposed by Sun Microsystems (Arnold et al. 2000). JavaScript is used to conduct scripting (simpler) programming for manipulating objects within a web browser. In addition to Java and JavaScript, ArcGIS has recently added Python to its list of programming languages (Van Rossum 2007).

Why do we need GIS programming? Mature GIS software and application templates provide many tools to accomplish our daily tasks; however, in order to understand the fundamentals of how GIS works and to customize software for specific problems, programming is required. The following list gives programming examples:

- *Customizing software for application*: The National Park Service is developing a simple web mapping application to allow the general public to interactively select and view information for a particular National Park. Using an online mapping tool such as Google Maps and selecting a park with your mouse will trigger a query of the selected information for that park. In this scenario, we need geographic information about the parks, a program for front-end user interaction, and a database query language that will generate result for the selected park.

- *Automating a process*: Suppose there are 100 geographic datasets collected in text file format and we need to convert them into a shapefile, a native data file format used by ArcView and ArcGIS, for further processing. ArcGIS can perform the conversion one by one, but doing this manually 100 times is monotonous. Therefore, a simple scripting tool to automatically read and process the 100 datasets into shapefiles would be beneficial. Using Python scripts in ArcGIS provides the capability to do so.

- *Satisfying simple GIS needs*: Suppose there is a transportation company that needs to track their company vehicles' positions based on 5-minute intervals. However, the company cannot afford to purchase a professional GIS software license. To resolve the issue, the company can use Python to create a map to show the company's service region and vehicle locations every 5 minutes. This programming may include Zoom In/Out, and Move/Pan features, animations based on locations, and a selection of one or many vehicles.

- *Cultivating advanced GIS professionals*: Suppose a group of students are asked to invent a routing algorithm based on predicted traffic conditions and real-time traffic information. The students will need to organize the road network information comparing real-time and predicted network speed. It is essential to use the most accurate predicted information in the routing process. Programming is needed throughout the entire process for network management and routing, and for reconstructing the results into map form or written directions.

Geographic information has become increasingly important in all walks of human life, whether it is for scientific discovery, forecasting natural disasters, advancing technologies of observations, or creating public awareness about location and routing. While some applications require complete GIS technologies to produce valuable results, many geographic information applications do not require sophisticated geographic information systems. For the latter case, open-source or small geospatial information software is utilized, while commercial GIS systems such as ArcGIS, are available for the former case. To better address both needs, it is essential to understand the fundamentals of how GIS works and its basic geographic information processing. This chapter introduces the background structure for building such capabilities: computer hardware and software, GIS and programming, GIS data models and Unified Markup Language (UML, Fowler 2004), and Python. Hands-on programming experience is needed for understanding the concepts and developing the essential skills utilized by GIS professionals in their work and research. Based on GIS fundamentals, this book will help you develop and improve systematic programming skills and will provide a more in-depth understanding of GIS fundamentals. Owing to its popularity within the GIS community, Python will be the primary programming language used in this book.

## 1.3 Python

Python was originally developed by a Dutch programmer, Guido van Rossum, in 1990. Van Rossum was reportedly a fan of the British comedy series, *Monty Python's Flying Circus*, and upon developing the open-source programming language, he borrowed to the name "Python" for the language and his nonprofit institution, the Python Software Foundation.

Similar to programming languages C++ and Java, Python is an object-oriented and interactive language. Python is dynamic in that it uses an automatic memory management mechanism to allocate and release memory for data (variables). Python and ArcGIS regularly release new versions of their programs; this book is based on Python release 2.7 and ArcGIS 10.1.

There are many reasons for choosing Python, including the following:[*]

- It is excellent for programming beginners, yet superb for experts.
- The syntax of Python is very simple and easy to learn. When you become familiar with them, you will feel that it is really very handy.
- It is highly scalable and well suited for both large and small projects.
- It is in a rapid development phase. Almost every half year, there is a new major release.
- It is portable cross-platform. This means that a program written in Windows can be run using the Linux or Mac operating systems.
- It is easily extensible. You can always add more class functions to your current project.
- It has powerful standard libraries.
- Many third parties also provide highly functional packages for you to utilize. Instead of developing GIS functions from scratch, you can simply download the source code and integrate them into your project.
- It is a fully *object-oriented* language, simple yet elegant, and stable and mature.

There are several steps to learning Python for GIS programming:

- Get familiar with the concept of class and object (Chapters 1 and 2).
- Learn the syntax of Python, including variables, data types, structures, controls, statements, and other programming structures (Chapters 1 through 4).
- Build Python programs from scratch and integrate open-source libraries to facilitate programming (Chapter 5).
- Become comfortable with the Python programming environment (Python interpreter or Python Text editor, Chapter 6).
- Solve GIS problems by writing code for GIS algorithms (Chapters 7 through 13).

These components are introduced in the above order throughout this book. This chapter introduces important concepts such as object-oriented programming, UML, and GIS models.

---

[*] http://pythoncard.sourceforge.net/what_is_python.html.

## 1.4 Class and Object

Within this section, we will discuss two types of fundamental concepts: class and object (Rumbaugh et al. 1991). Class uses a set of attributes and behaviors to represent a category of real-world phenomena. For example, Figure 1.3 shows how to extract the student attributes and behaviors.

Another example is online shopping on Amazon or eBay. Both the customers and online products must be abstracted into classes:

- Customers would have a customer ID, shipping address, and billing address. Customer behavior would include adding or deleting a product to the shopping cart.
- Products would have a product ID, product name, and product price. Product behavior would include setting the price, and totaling the product quantity/amount.

An object is a specific instance of a class. We can consider objects as instances of classes by assigning values to their attributes. Specifically, a class is the abstraction of a category or collection of real-world entities while an object is a specific real-world entity within the class. Within a computer, a class is the template and an object is the specific entity that occupies the computer memory. The computer can operate on both the attributes and behaviors of an object. For example, when a student logs in to their college web system with a username and password, the system will create a new student object. The computer reads each student as an independent object with several different attributes (e.g., username, password, and student ID). After logging in, a student is able to search, register, or add/drop classes using the object in the system, which represents him or her specifically. Chapter 2 will introduce how to define classes and objects using Python.

- Attributes
  - Student Id (G#)
  - Name
  - Email
  - Address
  - Courses
- Behaviors
  - register()
  - add_class()
  - drop_class()
  - change_address()

Student

Attributes: A student will have a set of attributes, including student ID, name, email, current home address, courses the student have registered for each semester.

Behaviors: Behaviors could include register the current semester, change address if the student moves to a different place, add or drop a class.

**FIGURE 1.3**
An example of representing students with the Student class.

## 1.5 GIS Data Models

GIS data models are used to capture essential geospatial elements of a specific problem (Longley et al. 2001). There are three types of data models: vector data, raster data, and special data. Vector data models consist of point, polyline, and polygon model types. Raster data includes equally split cells of digital elevation models and images. Special data are composed of network and linear data. This book highlights different types of GIS data models, but will focus mainly on vector data models.

A point can refer to a class of vector data represented by a pair of x, y coordinates in a two-dimensional (2D) space or a tuple of x, y, and z coordinates in a three-dimensional (3D) space. For example, a city is represented as a point on a world map. Each city has a group of attributes, which would include the city name, population, average household income, and acro-names. Another example using points is a map depicting all the restaurants within a certain region. In addition to its point location, each restaurant will contain other relevant information, including its name, room capacity, cuisine, and the year it opened. In these cases, the point is a general classification, whereas the city or the restaurant is a more specific type of class containing different attributes. When designing, each point of the rectangle can represent a class (Figure 1.4). This diagram is also referred to as a UML class diagram. The first row refers to the name of the class: *City*; the second row refers to the attributes of the class: *name* and *averageIncome*; the third row refers to a set of methods: *getName, getAverageIncome*, and *setName*.

Polylines are a class of vector data represented by a list of points. For instance, a river can be represented as a polyline on a map, which then can be categorized as a type of polyline class. A polyline class may include point coordinates, relevant attributes, and a set of methods. Another polyline dataset example can be roads, highways, and interstates. Both examples are categories of polylines. Rivers can be represented using UML (Figure 1.5). The first row of the UML is the subject of the class: *River*; the second row includes the river's attributes: *name* and *coordinates*; and the third row refers to the methods the programmer will use: *getName, setCoordinates*, and *setName*.

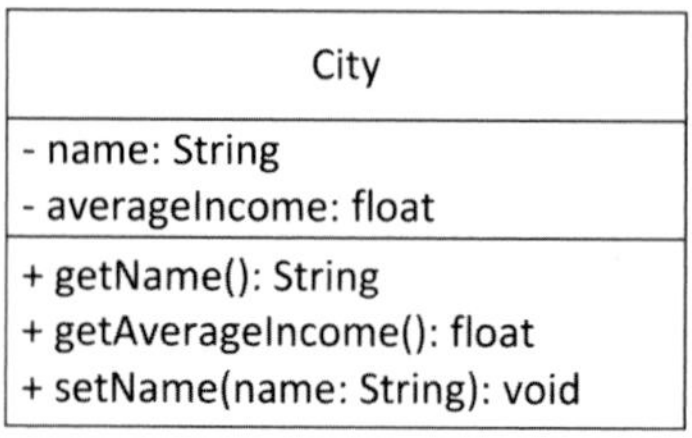

| City |
| --- |
| - name: String<br>- averageIncome: float |
| + getName(): String<br>+ getAverageIncome(): float<br>+ setName(name: String): void |

**FIGURE 1.4**
A UML diagram for the City class.

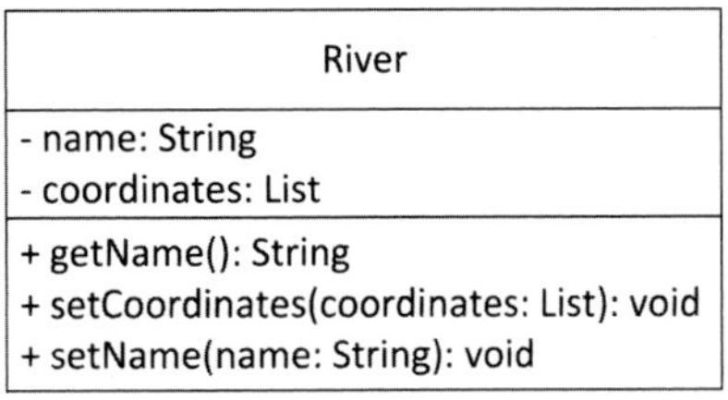

| River |
|---|
| - name: String |
| - coordinates: List |
| + getName(): String |
| + setCoordinates(coordinates: List): void |
| + setName(name: String): void |

**FIGURE 1.5**
The River class includes three parts.

| County |
|---|
| - name: String |
| - population: int |
| + getName(): String |
| + setPopulation(population: int): void |
| + setName(name: String): void |

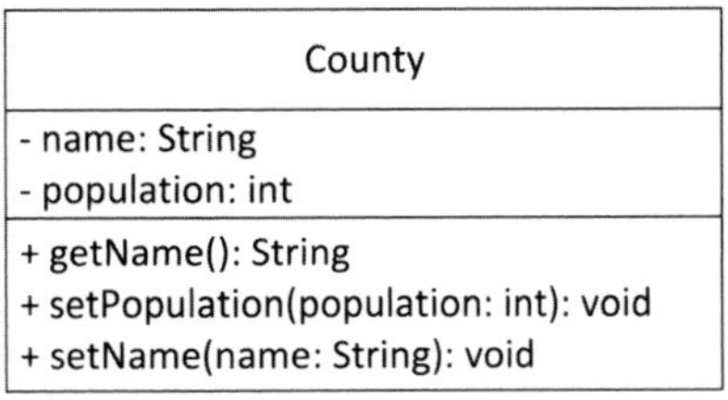

**FIGURE 1.6**
The County class includes three parts.

Polygons are another class of vector data that are also represented by a list of points; however, with polygons, the first and last points are the same. For example, on the map of the state of Virginia, a specific county, like Fairfax County, can be represented as a polygon. The county is a type of polygon class, which includes a list of points, relevant attributes, and a set of methods. Countries on a world map may also be represented as polygons. In either case, both the county and country are types of polygons. As shown in Figure 1.6, the first row is the subject name: *County*; the second row is the subject's attributes: *name* and *population*; and the third row refers to the methods: *getName, setPopulation,* and *setName*.

Developing more methods will require adding more methods and attributes to each class to capture the evolution of the data models and the functionality of software; UML diagrams are used to standardize their representation. This section uses class diagrams and relevant UML standards for the point, polyline, and polygon classes.

## 1.6 UML

In 1997, the Object Management Group (OMG)* created the UML to record the software design for programming. Software designers and programmers

* See OMG at http://www.omg.org/.

use UML to communicate and share the design. Similar to the English language in which we communicate through sharing our ideas via talking or writing, UML is used for modeling an application or problem in an object-oriented fashion. UML modeling can be used to facilitate the entire design and development of software.

The UML diagram is used to capture the programming logic. There are two types of diagrams that we will specifically discuss: class diagrams and object diagrams (Figure 1.7).

The UML class diagram can represent a class using three parts: name, attributes, and methods. The attributes and methods have three different accessibilities: public (+), private (-), and protected (#). Attributes and methods are normally represented in the following format:

- Attributes: accessibility attribute Name: Attribute data type, for example, +name: String
- Methods: accessibility method Name (method arguments): method return type, for example, +setName(name:String): void

Public refers to the method/attributes that can be accessed by other classes. Private methods/attributes cannot be accessed by any other classes.

Protected methods/attributes cannot be accessed by other classes except those classes inherited from this class (explained below).

There are several fundamental relationships among different classes: dependency, inheritance, composition, and aggregation. Dependency represents one class dependent on another. Inheritance is an important relationship in which a class is a subtype of another class. Figure 1.8 illustrates the dependency between geometry and coordinate systems in that the existence of geometry depends on a coordinate system. This relationship is represented by a dashed line and an arrow from the geometry to the coordinate system class. The relationship between a point, line, and polygon are classified within the geometry class.

Aggregation and composition are two other important relationships in UML. Aggregation represents "has a" relationship in UML. For example, a state is an aggregation of a number of counties (Figure 1.9a). Composition represents, or "owns" relationship. For example, a multipoint class may be composed of two or more points (Figure 1.9b).

The relationship can be quantified by the number of elements involved. For example, a line includes 2+ points and a state includes 0+ counties. There are six different types of this multiplicity relationship (Figure 1.10). A multipoint is composed of two or more points (Figure 1.9b) and a state is aggregated by zero or more counties.

An object is an *instantiation* of a class. The object diagram shows a complete or partial view of the model system structure at a specific time. So, the state

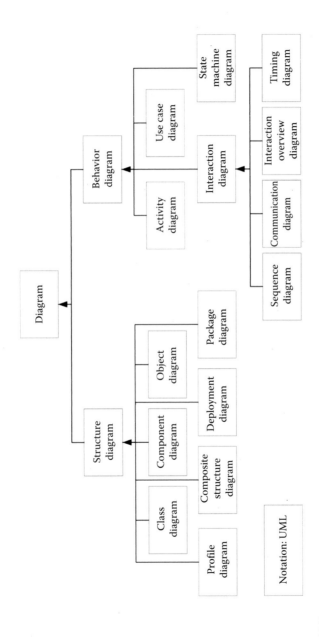

**FIGURE 1.7**
The class diagram and object diagram used in this book.

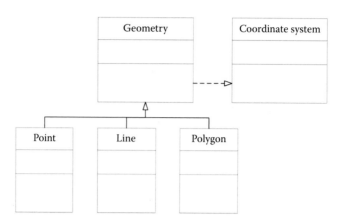

**FIGURE 1.8**
Inheritance and dependency.

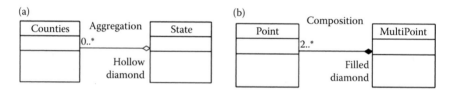

**FIGURE 1.9**
(a) Aggregation and (b) composition are two polar relationships among classes.

```
0..0: empty collection
0..1: 0 or 1
0..* or *: from 0 to many
1..*: from 1 to many
m..n: from m to n
1: one and only one
```

**FIGURE 1.10**
Multicity relationship among classes.

of an object can be changed. Figure 1.11's class name is worldMap, and its object is the coordinate system that changed from WGS 1972 to WGS 1984 after performing reprojection.

## 1.7 Hands-On Experience with Python

A point is the basic data model within GIS. This section will examine how to create a point class, including coordinates and calculations of the

| worldMap: Map |
| --- |
| coordinateSystem = "WGS 1972"<br>Date = "08/29/03"<br>Scale = "1:500" |

| worldMap: Map |
| --- |
| coordinateSystem = "WGS 1984"<br>Date = "08/29/03"<br>Scale = "1:500" |

Initially, coordinateSystem = "WGS 1972", and after performing reprojection, coordinateSystem = "WGS 1984"

**FIGURE 1.11**
worldMap is an object of the Map class and the state is changing with different operations.

distances between points. You will learn how to create point objects from point class.

1. Open the program (Figure 1.12):

   Windows→All Programs→ArcGIS→Python 2.7

   or

   Windows→All Programs→Python 2.7→IDLE (Python GUI)

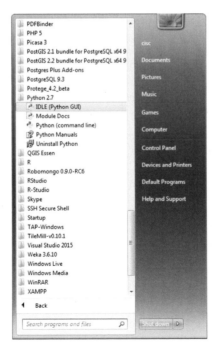

**FIGURE 1.12**
Launch the Python programming window (GUI).

```
>>> import math
>>> class Point():
        def __init__(self):
                self.x = 0
                self.y = 0
        def setXY(self,x,y):
                self.x = x
                self.y = y
        def calDis(self,p):
                return math.sqrt((self.x-p.x)**2+(self.y-p.y)**2)

>>> p1 = Point()
>>> p2 = Point()
>>> p1.setXY(1,2)
>>> p2.setXY(2,3)
>>> p1.calDis(p2)
1.4142135623730951
>>>
```

**CODE 1.1**
Creating a point class and generating two points, then calculating the distance between the two points.

2. Type in the point class codes as shown in Code 1.1.

Programming tips:

1. Coding should be exactly the same as the figure shows.
2. The init method is defined with four underscores: two "_" before and two after "init."
3. Python is case sensitive, so lower- and uppercase of the same letter will make a difference.
4. There is no need to understand every statement for now; they will be gradually explained in the following chapters.

## 1.8 Chapter Summary

This chapter briefly introduced GIS programming and included

- A general introduction to computer hardware and software
- Definitions of GIS and programming
- Python in a practical context
- Practical knowledge about several GIS data models

- The unified modeling language for modeling object-oriented GIS data
- Relevant hands-on experience

**PROBLEMS**

- Define computer, programming, software, and GIS.
- What are the different methods to categorize software?
- What are the three GIS data models found on the UML diagram?
- Explain why we need to learn GIS programming.
- Use the UML diagram to model the relationship between polylines.
- Use the UML diagram to model the relationship between polygons.
- Practice Python's Chapter 3 tutorial: https://docs.python.org/3/tutorial/introduction.html.
- Use Python to calculate the distance between Point (1, 2) and Point (2, 2).
- Discuss how to identify classes used on a world map and how to use UML to capture those classes.

# 2

## Object-Oriented Programming

This chapter introduces object-oriented programming in regard to Python's programming language, classes and objects, object generation, inheritance, GIS classes and objects, and a general programming experience.

### 2.1 Programming Language and Python

Programming language is defined as an artificial language used to write instructions that can be translated into machine language and then executed by a computer. This definition includes four important aspects: (1) artificial language, a type of programming language created solely for computer communication; (2) instruction based, a programming language with limited instructions supported by a specific computer or CPU; (3) translation, the conversion from human instructions to a technical computer program, or CPU; and (4) translator, of which there are two types: interpreter and compiler (Aho and Ullman 1972).

There are two different methods computer programmers use to convert languages into a legible format on the computer. One method requires a computer programmer to compile a group of statements written in a specific language and convert them into a machine-readable format prior to running the program. The other method entails simultaneously translating the language while running the program. For example, in C programming, we need to use C compiler to translate the program into machine codes before execution. Similarly, C++ and Java are compiling-type programing languages. BASIC programming language is an interpreter language (Lien 1981), in which the interpreter will translate the program while it is running. Likewise, Python, Perl, and PHP are considered interpreter languages. Therefore, in order to successfully use Python on a computer, a Python interpreter must also be installed.

Programming languages have evolved considerably from machine and assembly languages to intermediate and advanced languages (Rawen 2016). Machine language instructions are represented in a specific sequence using 0s and 1s. One single digit, or number, is called a bit. A combination of three bits is called an octal number (an eight digit combination using the numbers 0–7), whereas a combination of four bits is called a hex number

English:
>           Print 'A' for 1000 times.

Python:

```
for i in range(1000):
        print 'A'
```

C:

```
for (i=0; i<1000; i++) printf("A");
```

Assembly language:

```
0360 A9 01 LDA #$01
0362 A0 00 LDY #$00
0364 99  00 80 STA $8000,Y
0367 99  00 81 STA $8100,Y
036A 99 00 82 STA $8200,Y
036D 99 00 83 STA $8300,Y
0370 C8 INY
0371 D0 F1 BNE $0364
0373 60 RTS
```

Sample Machine Language:

```
169 1 160 0 153 0 128 153 0 129 153 130 153 0 131 200 208 241 96
```

**FIGURE 2.1**
Print 'A' 1000 times using different types of languages.

(a 16 digit combination using the numbers 0–15). Assembly languages depict the machine bit operations with easy-to-remember text representations. Intermediate languages are typically more powerful and easy to code. Advanced languages are more similar to human language, do not have access to specific hardware functions, and are executed on several different hardware types.

The example uses different representations for the "print letter 'A' for 1000 times" (Figure 2.1).

Machine languages become increasingly difficult to understand by humans, so only specific CPUs are able to read the language accurately (Hutchins 1986). Therefore, in GIS programming, we normally use advanced languages such as C, Java, or Python instead of machine or assembly language.

C is a typical procedural language that was developed around 1969–1973 and became available to the general public around 1977–1979. It was officially standardized by the ANSI X3J11 committee in the mid-1980s and has become one of the most commonly used languages in the computer industry. The early editions of GRASS (Geographic Resource Analysis Support System, Neteler and Mitasova 2013) GIS* open-source software and ArcGIS were developed using C. Bjarne Stroustrup of Bell Laboratories expanded C to C++ in order to support object-oriented features. C++ supports C features in function calls and object-oriented classes/objects fashion. Both C and C++ are complex for beginning programmers. Since 1999, ISO/ANSI has

---

* http://grass.osgeo.org/.

standardized C++ to improve and maintain state-of-the-art quality within the industry. C and C++ are commonly used in Linux and have influenced other languages such as C# and Java.

Developed by Sun at SunWorld'95, Java is a pure object-oriented language developed to target Internet and cross-platform applications. Over time, Java has become increasingly popular among IT companies such as Microsoft, Borland/Eclipse, IBM, and Sun. The official Java resource can be found at java.sun.com and an open-source compiler/programming environment can be found on the Eclipse Foundation website at www.eclipse.com.

Python is an interactive language programming system created by Guido van Rossum in 1990. Python is dynamically written and uses automatic memory management. The nonprofit Python Software Foundation consistently updates and manages this open-source project. Python is fully developed in that it can write once and run many times on different platforms. This book will analyze and explain Python as it is applied to GIS and ArcGIS* programming. You can download any version from Python's website; however, not all versions interactively work with ArcGIS. Python is easy to learn and use, and is supported by ArcGIS, which is why we have chosen it to be the programming language for this book.

## 2.2 Class and Object

*Classes* and *objects* are widely used in Python. Class defines the template for a category of objects with name, attributes, and methods. Objects are instances of classes with attributes and methods. The attributes and methods can be referred to using a '.'. For example, the coordinate attributes and calDis method of a point object created from a Point class can be referred to using point.x, point.y, and point.calDis().

### 2.2.1 Defining Classes

Python provides the mechanism to define a class using the keyword class with the syntax of 'class className:', for example, 'class Point:', 'class Polyline:', or 'class Polygon:'. The attributes and methods can be defined for a class using the 'def' keyword. Figure 2.2 shows the Python code for defining a Point class with attributes name, x, y defined and the method setName() defined. In the __init__ method, "0, 0" was passed in as value for x, y, and name.

Many classes define a special method named __init__() to create/construct objects. The method will be called when we create an object using the class

---

* http://www.esri.com/software/arcgis.

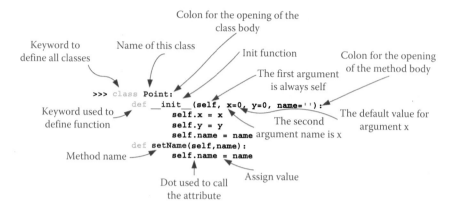

**FIGURE 2.2**
An example of defining a Point class with Python.

(such as Point class here). The __init__ method has four '_'—two before and two after 'init'—to make it the construction method that will be used when creating an object. For all methods defined by a class, the first parameter is always 'self', which refers to the object itself. This can be used to refer to the attributes and methods of the objects. For example, the __init__ method will create a point object with self as the first parameter and x, y, name initial values for the object. By default (without specifying the values), the values for x, y, and name will be 0, 0, and blank string, respectively. The first two statements of Code 2.1 create two point objects (point0 and point1). The object point0 is created with default values and point1 is created with arguments

```
>>> class Point:
        def __init__ (self, x=0, y=0, name=''):
                self.x = x
                self.y = y
                self.name = name
        def setName(self,name):
                self.name = name


>>> point0 = Point()
>>> point1 = Point(1,1,'first point')
>>> point0.x, point0.y, point0.name
(0, 0, '')
>>> point1.x, point1.y, point1.name
(1, 1, 'first point')
>>> point1.setName('second point')
>>> point1.name
'second point'
>>>
```

**CODE 2.1**
Creating a point may pass in value to the object through parameters.

of 1, 1, and 'first point'. If no parameters are given when creating point0, the default values 0, 0, and ' ' will be used. When values (1, 1, 'first point') are given parameters, the __init__ method will assign the values passed into the attributes of point1.

## 2.2.2 Object Generation

To create an object, type **objectName** = **className()** with none or multiple parameters, which will be passed to the attributes declared in the __init__() methods.

*objectName = className(value1,value2,...)*

In Code 2.1, we generated two objects, point0 and point1. While declaring object point0, no parameter is passed while three values (1, 1, 'first point') are used to generate point1.

To refer to an object's attribute or method, we start with the *objectName*, followed by a period and then end with the attribute name or method name.

*objectName.attributeName*

*objectName.methodName()*

Code 2.1 uses .x, .y, and .name following the objects point0 and point1 to refer to the attributes x, y, and name. The instruction *point1.setName()* is called to change the name of point1 to 'second point'.

## 2.2.3 Attributes

Each class may have one or more attributes. Section 1.4 explains how attributes can be public, private, or protected to indicate different accessibility by other objects. How do you explicitly specify the public and private attributes while declaring a class?

- *Public:* Attributes in Python are, by default, "public" all the time.
- *Private:* Attributes that begin with a double underscore ("_"). Such attributes can be protected as private because it cannot be directly accessed. However, they can be accessed by *object._ClassName_attributeName*, for example, test._Test_foobar, where test is an object of Test class, and _foobar is a private attribute (Code 2.2).
- *Protect:* Attributes prefix with a single underscore "_" by convention. However, they still can be accessed outside of the class in Python.

Another important attribute in Python is the static attribute, which is used to hold data that is persistent and independent of any object of the class

```
>>> class Test:
        def __init__(self):
                self.__foobar = "private attr"
                self.foobar = "public attr"

>>> test = Test()
>>> test.foobar
'public attr'
>>> test.__foobar

Traceback (most recent call last):
  File "<pyshell#23>", line 1, in <module>
    test.__foobar
AttributeError: Test instance has no attribute '__foobar'
>>> test._Test__foobar
'private attr'
>>>
```

**CODE 2.2**
Declare public, private, and protect attributes.

(Code 2.3). For example, we can create a map including different layers, and the layer scale can be static and the same to all layer objects.

A class (and instantiated object) can have special built-in attributes. The special class attributes include a class name and description of the class (Code 2.4).

```
>>> class Test:
        version = 1.0


>>> Test.version
1.0
>>> t1 = Test()
>>> t2 = Test()
>>> t1.version
1.0
>>> t2.version
1.0
>>> Test.version = 2.0
>>> t1.version
2.0
>>> t2.version
2.0
>>> t1.version = 3.0
>>> t1.version
3.0
>>> Test.version
2.0
>>> t2.version
2.0
>>>
```

**CODE 2.3**
Declare static attributes.

```
>>> class Point:
        """Point Class Definition"""
        def __init__(self):
                self.x = 0.0
                self.y = 0.0
        def getDistance():
                pass ## ignore here


>>> Point.__name__
'Point'
>>> Point.__doc__
'Point Class Definition'
>>> Point.__module__
'__main__'
>>>
```

**CODE 2.4**
Special class attributes.

```
>>> p1 = Point()
>>> p1.__class__
<class __main__.Point at 0x02A100D8>
>>> p1.__dict__
 {'y': 0.0, 'x': 0.0}
>>>
```

**CODE 2.5**
Special object attributes.

- _name_: class name
- _doc_: description
- _bases_: parent classes
- _dict_: attributes
- _module_: module where class is defined

The special object attributes include a class name and an object's attributes (Code 2.5).

- _class_: class from which object is instantiated
- _dict_: attributes of object

## 2.2.4 Inheritance

Chapter 1 introduces three important relationships among objects in object-oriented programming: inheritance, encapsulation, and polymorphism. Inheritance is an efficient way to help reuse a developed class. While private attributes and methods cannot be inherited, all other public and protected attributes and methods can be automatically inherited by subclasses.

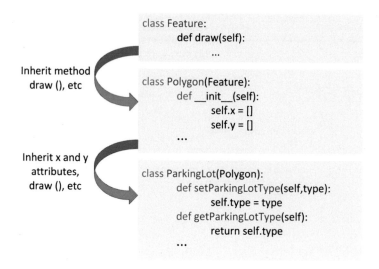

**FIGURE 2.3**
An example of inheritance (*ParkingLot* class inherits from class *Polygon*, and *Polygon* inherits from *Feature*).

To inherit a super class in Python, include the super class name in a pair of parentheses after the class name.

*class DerivedClassName(SuperClass1)*

We can also inherit multiple classes in Python by entering more than one class name in the parentheses.

*class DerivedClassName(SuperClass1, SuperClass2, SuperClass3)*

Figure 2.3 shows an example of inheritance. Assuming we have a class *Feature*, which includes a method *draw()*, then the class Polygon will inherit from the class *Feature*. With this inheritance, the Polygon class will have the method *draw()* as well. When we define the *ParkingLot* class with the inheritance from the *Polygon*, the *ParkingLot* will have attributes of x and y coordinates as well as the method *draw()*. The *Polygon* and *ParkingLot* may have different drawing implementations; however, you can use the *draw()* feature for both the *Polygon* and *ParkingLot*. This particular method using different implementations for different subclasses is called polymorphism.

## 2.2.5 Composition

Composition is an efficient way to help us reuse created objects, and to maintain the part-to-whole relationship between objects. To maintain the

```
class Feature:
        def draw(self):
                ...

class Point(Feature):
        def __init__(self):
                self.x = []
                self.y = []
        ...

class Polygon(Feature):
        def __init__(self, points):
                self.points = points
        ...
```

**FIGURE 2.4**
Composition example (a *Polygon* class includes attribute points as objects generated from class *Point*).

composition relationship, you must define a class with an attribute that can include a number of other class objects.

Figure 2.4 shows an example of composition. The class *Point* and the class *Polygon* inherit from the class *Feature*. The class *Polygon* border is defined by a sequence of points formed in a ring and is captured by point attributes. The points' coordinates are kept in the point objects. Not only does this show how a *Polygon* object requires a number of *Point* objects, but also the composition relationship between *Point* and *Polygon*.

## 2.3 Point, Polyline, and Polygon

In GIS, there are three basic vector data types, which include Point, Polyline, and Polygon (Chang 2006). We can abstract those features and define a class for each of them to be reused. We can also define a Feature class as a super class for Point, Polyline, and Polygon. The following are examples of the four classes:

- *Feature:* Normally, a *Feature* class (Figure 2.5) has a name to keep the feature name and a method *draw()* to draw the feature on a map. The draw method should include at least two parameters, self and map. Self refers to the object accessing feature object data while drawing, whereas a map refers to the background that we will draw the feature on.

| Feature |
| --- |
| + name: string |
| + draw(a: Feature, b: Map): void |

**FIGURE 2.5**
UML design for *Feature* class to define common attributes and methods of *Point, Polyline,* and *Polygon.*

```
>>> class Feature:
        def __init__(self,name = ''):
            self.name = name
        def draw(self,map):
            pass

>>>
```

**CODE 2.6**
Define a Feature class as a super class for Point, Polyline, and Polygon.

| Point(Feature) |
| --- |
| +x: float |
| +y: float |
| + calDis(a: Point, b: Point): float |

**FIGURE 2.6**
UML design for *Point* class to keep point vector data.

For example, Code 2.6 is an example of defining *Feature* as a class in Python:

- *Point:* A Point class (Figure 2.6) should have at least two spatial attributes, x and y, to represent the coordinates of a point. Another non-spatial attribute could include the name. A Point class could also use *calDis()* to calculate the distance between two points. The argument for the *calDis* method is current point object "self," and another point object "point." The return value for the distance between two points is designated as float.

For example, Code 2.7 is an example of defining Point as a class in Python:

After declaring a Point class, we can initiate objects. For example, we can populate two points with data (1, 2), (2, 2), and then calculate the distance between the two points (Code 2.8).

- *Polyline:* Defining a polyline class requires different attributes to keep the data of polylines and methods by polylines. For example, two lists of x, y coordinates or a list of points can be used to keep

```
>>> import math
>>> class Point(Feature):
        def __init__(self,x =0.0,y = 0.0):
                self.x = x
                self.y = y
        def calDis(self,point):
                return math.sqrt((self.x-point.x)**2+(self.y-point.y)**2)

>>>
```

**CODE 2.7**
Define a Point class in Python.

```
>>> p1 = Point(1,2)
>>> p2 = Point(2,2)
>>> p1.calDis(p2)
1.0
>>>
```

**CODE 2.8**
Calculate the distance between (1, 2) and (2, 2).

the x, y coordinates depending on the design. For object-oriented purposes, we can use the second approach (Figure 2.7a). For better system performance, data points in polylines are different from real objects in GIS, so we use the first approach (Figure 2.7b).

- *Polygon:* A *Polygon* class (Figure 2.8) could have one attribute, "points," to represent the list of all points used to define the border or two lists—the x and y coordinates for all such points. A *Polygon* class may also have a method *getLength()* to calculate the border

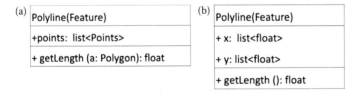

| (a) Polyline(Feature) |
| --- |
| +points: list<Points> |
| + getLength (a: Polygon): float |

| (b) Polyline(Feature) |
| --- |
| + x: list<float> |
| + y: list<float> |
| + getLength (): float |

**FIGURE 2.7**
(a) UML Polyline class uses point object list to keep coordinates for polylines. (b) UML Polylines class uses x and y lists to keep coordinates data for Polylines.

| Polygon(Feature) |
| --- |
| +points: list<Points> |
| +getLength (): float |

**FIGURE 2.8**
UML design for *Polygon* class to keep polygon vector data.

```
>>> # import math module from Python
>>> import math
>>> class Polyline(Feature):
        #Initiate method for a polyline object
        def __init__(self, points = []):
            self.points = points
        #A method to set points
        def setPoints(self, points):
            self.points = points
        #A method to get the length of the polyline object
        def getLength(self):
            length = 0.0
            for i in range(len(self.points)-1):
                length+=math.sqrt((points[i].x-points[i+1].x)**2+
                                    (points[i].y-points[i+1].y)**2)
            return length

>>>
```

**CODE 2.9**
A Polyline class has one attribute (points), two methods (*setPoints()*, and *getLength()*).

length of the Polygon without arguments. The return value for the border length of polygon is designated as float.

## 2.4  Hands-On Experience with Python

The Code 2.9 defines a Polyline class and creates a polyline object. Type the example onto your computer and describe what each line defines and why it is included in the code.

## 2.5  Chapter Summary

This chapter discusses object-oriented programming as well as how to program objects, classes, and relationships. After reading, you should be able to do the following:

- Understand object-oriented programming using Python.
- Define a class and know how to create objects using Point, Polyline, and Polygon.
- Practice inheriting super classes.
- Know how to reuse the code and its necessary functions.

## PROBLEMS

1. Pick three points, for example, (1, 100), (25, 60), and (1, 1). Could you form a polyline or polygon using these three points?

2. Create an algorithm to calculate the distance between two points, for example, $(x_1, y_1)$, $(x_2, y_2)$.

3. Read Python Tutorial 6.2 and 6.3. (Use Python command line window for 6.2).

4. Define and program three classes for *Point*, *Polyline*, and *Polygon*.

5. Add distance calculation in-between every two points, and program to calculate the distance among the three points given.

6. Add the *getLength()* method in *Polyline* and *Polygon*; create a polyline and polygon using the three points given; calculate the length of the polyline and perimeter of the polygon.

# Section II

# Python Programming

# 3

## Introduction to Python

Learning a programming language is a practical and progressive journey. You will begin with the basics, and gradually increase your skills at complex coding. This chapter will introduce fundamental Python components, including classes and objects, syntax, data types, operators, functions, flow control, exceptions, input/output, and modules. A number of libraries are available for specific development, including graphical user interfaces, databases, Internet, and web programming. To facilitate the learning process, you will utilize your understanding of GIS to start building a mini-GIS package while learning the programming language.

## 3.1 Object-Oriented Support

One of Python's most important characteristics is object-oriented structure. Foundational Python programming requires understanding concepts about classes and objects, inheritance, composition, package, and class sharing. Python provides the mechanism to define a class and create objects from that class.

*Classes and objects are widely used in Python.* Objects are instances of classes. Objects have attributes and methods. Dot(.) refers to attributes and methods of an object. In Chapter 2, this book defined *Point* class and created *p1* as a point object. You can use *p1.x* and *p1.calDis()* to call the attribute and method of the object. Inheritance helps you reuse the attributes and methods of a class (superclass) by using the super class to define new classes in the format 'class subclass (superclass)'. All public and protected attributes and methods are inherited by subclasses automatically. Private attributes and methods will not be inherited. For example, three vector classes (*Point, Polyline,* and *Polygon*) are defined by inheriting the Feature class's method *draw()* as detailed in Chapter 2. Composition helps maintain the part–whole relationship, where one object of a class includes objects of other classes. Python uses object reference to implement the composition relationship. As discussed in Chapter 2, a list of points are used to keep the coordinates of a polyline or polygon border.

When classes and methods are developed for software, similar classes/methods are grouped into the same module for easier organization. So, all

GIS data classes can be put into one big module. Submodules include classes/ methods for vector, raster, and special data types. If other classes need to access the classes/methods in a module, the module must be imported first. For example, the math module in Code 2.5 and Code 2.6 are imported to access all math methods (e.g., math.sqrt).

## 3.2 Syntax

### 3.2.1 Case Sensitivity

Python is *case sensitive*, meaning capital and lowercase letters represent different identifiers. You can define a variable myList with an uppercase *L*, and store the list of items "1, 2, 3, and 4." If you get the first value of the list using *mylist[0]* with a lowercase *l*, you will see a *NameError*, which shows that *mylist* is not defined because you defined myList using a capital L (Code 3.1).

### 3.2.2 Special Characters

Python has a list of special characters with special meanings. Table 3.1 lists some common special characters and their respective functions.

### 3.2.3 Indentation

In Python, indentation is important for grouping code. Indented lines start at different positions or column; numbers are not allowed, they will trigger an *IndentationError*. You may use a different number of spaces or columns to indent different levels of statements; however, 4 or 8 spaces are recommended. Therefore, *space and tabs* play significant roles in organizing codes. Different program editors (e.g., command line and Python GUI) use "tab" in different manners. Depending on your text editor, it may represent different numbers of spaces.

```
>>> myList = [1,2,3,4]
>>> myList[0]
1
>>> mylist[0]

Traceback (most recent call last):
  File "<pyshell#2>", line 1, in <module>
    mylist[0]
NameError: name 'mylist' is not defined
>>>
```

CODE 3.1
Case sensitive.

**TABLE 3.1**

Special Characters in Python

| Symbols | Function | Example |
|---|---|---|
| \ | Escape characters that have a special meaning | ```>>> print ''test''```<br>```test```<br>```>>> print '\'test\''```<br>```'test'```<br>```>>> print '\\test'```<br>```\test``` |
| \n | New line | ```>>> print 'first line\nsecond line'```<br>```first line```<br>```second line``` |
| \t | Tab | ```>>> print 'str1\tstr2'```<br>```str1    str2``` |
| : | Go to next level of statements | ```>>> class Polyline:```<br>```        def getLength():```<br>```            pass``` |
| # | Indicate Python comments | ```>>> # this is a comment``` |
| ; | Join multiple statements on a single line | ```>>> import math; x = math.pow(2,3)```<br>```>>> import math y = math.pow(2,3)```<br>```SyntaxError: invalid syntax``` |

## 3.2.4 Keywords

Keywords, such as def and del, are reserved words and cannot be used for any other purpose (e.g., as the names of variables, classes, and objects); otherwise, a *SyntaxError* will occur (Code 3.2). Table 3.2 lists the keywords in Python.

```
>>> x =and
SyntaxError: invalid syntax
>>>
```

**CODE 3.2**
Keywords SyntaxError example.

**TABLE 3.2**

Python Keywords

| | | | |
|---|---|---|---|
| and | elif | global | or |
| assert | else | if | pass |
| break | except | import | print |
| class | exec | in | raise |
| continue | finally | is | return |
| Def | for | lambda | try |
| Del | from | not | while |

```
>>> a = b = c = 0
>>> a1,b1,c1= 1, 1.0, 'c1'
>>> (a1,b1,c1)=(2, 2.0, 'c2')
>>> (a2,b2,c2)=(2, 2.0, 'c2')
>>> a, b, c, a1, b1, c1, a2, b2, c2
(0, 0, 0, 2, 2.0, 'c2', 2, 2.0, 'c2')
>>>
```

**CODE 3.3**
Multiple assignments.

### 3.2.5 Multiple Assignments

Multiple assignments are useful in declaring variables simultaneously.
In Code 3.3,

- The first line of code assigns the same value to multiple variables by using "*a = b = c = value*."
- The second and third lines of code assign different values to different variables by using "a1, b1, c1 = v1, v2, v3," or "(a2, b2, c2) = (v1, v2, v3)."

### 3.2.6 Namespace

*A namespace* is a place in which a name resides. Variables within a namespace are distinct from variables having the same names but located outside of the namespace. It is very easy to confuse names in different namespaces. Namespace layering is called scope. A name is placed within a namespace when that name is given a value. Use *dir* to show the available names within an indicated namespace. For example, *dir()* can find current namespace names, *dir(sys)*

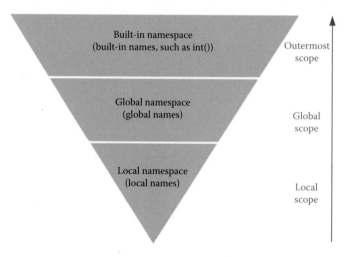

**FIGURE 3.1**
Hierarchy of namespaces.

will find all names available from *sys*, and *dir(math)* will find all names available from *math*. A program typically includes three layers of scope (Figure 3.1):

- The top layer is a system built-in namespace, which includes names defined within Python itself; these are always available in your programming environment.
- The middle layer is global namespace defined within a file or module.
- The bottom layer is local namespace, which includes names defined within a function or a class method.

## 3.2.7 Scope

*Scope* refers to a portion of code and is used to identify the effectiveness of variables. Scope is important in functions, modules, classes, objects, and returned data. Modules with function and data act similar to objects. For example, when defining *Point* class, use *p1* as a variable within the *calDis()* function of the class; or use *p1* to refer to an object later when creating a point object. The first *p1* is only effective in the scope of the *Point* class' *calDis()* function. The second *p1* is only effective at the same level as the overall program without indentation.

Variables can be local or global:

- Local variables are declared within a function, and are only accessible within the function. Once the function is executed, the variable will go out of scope.
- Global variables are nonlocal and can be accessible inside or outside of functions.

In the example of Figure 3.2, global_x is a global variable and local_y is a local variable. If you use local_y variable outside the function, you will get an error. For more information about namespaces and scope, please refer to Raschka (2014).

```
>>> global x = 5  ◄──────── Global Variable
>>> def add():
        local_y = 3  ◄──────── Local Variable
        return global_x + local_y
>>> print add()
8
>>> print local_y
Traceback (most recent call last):
    File "<pyshell#379>", line 1, in <module>
        print local_y  ◄──────── 'local_y' goes out of scope
NameError: name 'local_y' is not defined
```

**FIGURE 3.2**
Local and global variables.

## 3.3 Data Types

In python, *data types* are dynamic so that you do not have to define a variable's data type. The type of variable will be assigned when a value is assigned. The change in values will change the type of variables. There are two categories of data types: basic data types (integer, floating, long, complex, strings, type, and function) and composite data types (lists, tuples, and dictionaries).

### 3.3.1 Basic Data Types

The basic data types include number and string categories. The number category includes integers, long integers, and float (Table 3.3).

- Integers

  Integers are equivalent to integers in C programming language. The range of an integer is limited as follows:

  $-2^{31} \sim 2^{32}$ (–2147483648~4294967296)

  The integer value can be represented by a decimal, octal, and hexadecimal format. For example, *010* is an octal representation of 8 and *0x80* is a hexadecimal of 8.

- Long integers of nonlimited length

  The range of long integer is only limited by computer memory. A long integer is denoted by appending an upper- or lowercase "L." It can also be represented in decimal octal and hexadecimal formats.

- Float

  Floating numbers are equivalent to doubles in C language. A float value is denoted by a decimal point (.) in the appropriate place and

**TABLE 3.3**

Basic Data Types

| Basic Variable Types | | Range | Description | Examples | Conversion (Typecast) |
|---|---|---|---|---|---|
| Number | Integer | $-2^{31} \sim 2^{32}$ | decimal , octal, and hexadecimal format | 20, –20, 010, 0x80 | int(), e.g., int(2.0), int('2'), int(2L) |
| | Long integer | Limited only by memory | Denoted by (L) or (l). | 20L, –20L, 010L, 0x80L | long(), e.g., long(2), long('2') |
| | float | Depends on machine architecture and python interpreter | Denoted by a decimal point (.) | 0.0, –77.0, 1.6, 2.3e25, 4.3e-2 | float(), e.g., float(2), |
| String | | N/A | Denoted by ("), ("") | 'test', "test" | str(), e.g., str(2.0) |

```
>>> x = int (2.01)
>>> x
2
>>> float(2)
2.0
>>>
```

**CODE 3.4**
Data type conversion.

an optional "e" suffix (either lowercase or uppercase) representing scientific notation. The precision, or range of the float, depends
on the architecture of a machine as well as the Python interpreter
used.

• Conversion of numbers

The *int(), long(), float()* built-in functions are used to convert from any
numeric type to another (Code 3.4).

*Tips*

*Typecast:* Convert data from one type to another, for example, float ('3.14'),
which casts a string data type to float.

Type conversed assignment may result in lost precision, for example

y = 3.14
x = int(y)

where x will lose the precision values and has a value of 3 as a result.

• Strings

String data are denoted by single quotes " or double quotes "".

• Other built-in types

There are several other built-in data types, such as type, None, function, and file. Code 3.5 illustrates the following types:

*Function type ()* takes an object as an argument and returns the
data type of the object.

*None* is the null object and has no attribute.

*bool* object has two potential values: *True* and *False*. Conditional
expressions will result in Boolean value as either *True* or *False*.

*Tips*

Different from C or Java language, Python does not support Byte, Boolean,
Char, Pointer data types.

```
>>> x = type(1)      >>> x = type(True)
>>> x                >>> x
<type 'int'>         <type 'bool'>
>>> x == int         >>> x == int         Conditional expression
True                 False

>>> type(None)
<type 'NoneType'>
>>>
```

**CODE 3.5**
Function type, None type, and bool type.

### 3.3.2 Composite Data Types

Another category of built-in variable data types is called a composite data type (Table 3.4). Composite data types are ordered and sequentially accessible via index offsets in the set. This group, also known as sequences, typically includes list, tuple, dictionary, and set. A list is different types of data separated by commas, and are included in a pair of brackets, for example, [1, 2, 3, 4]. Tuples are lists of different types of data, such as (1, "the first rated products," "milk") set off by parentheses. Dictionaries are lists of keys and value pairs, as shown in the following example {'john':'3-1111', 'phil':'3-4742', 'david':'3-1212'}

- List

  The most commonly used and important composite data type is *list*, which can be used to group different values together. Use *list* to keep a series of points, polylines, and polygons in GIS programs.

  **Define**: A list object can be created from: *[v1, v2, v3, ....]*, where elements are surrounded by a square bracket (e.g., as in Code 3.6).

  **Operators:** Composite data types are good for expressing complex operations in a single statement. Table 3.4 lists common operators shared by complex data types.

- seq[index]: gets a value for a specific element. The starting index of all sequence data is 0, and the end index is one fewer than the number of elements $n$ in the sequence (i.e., $n$-1) (Code 3.6a).

**TABLE 3.4**

Composite Data Types and Common Operators

| Container | Define | Feature | Examples | Common Operators |
|---|---|---|---|---|
| List | Delimited by [ ]; | mutable | ['a', 'b', 'c'] | seq[index], |
| Tuple | Denoted by parenthesis () | immutable | ('a', 'b', 'c') | seq[index1: index2], seq * expr, |
| dictionary | {key: value, key: value , ...} | mutable | {'Alice': '7039931234', 'Beth': '7033801235'} | seq1 + seq2, obj in seq, obj not in seq, len(seq) etc |

```
>>> a = [1,2,3,4]
>>> a
[1, 2, 3, 4]
>>> b = [x*3 for x in a]
>>> b
[3, 6, 9, 12]
>>> a[0]
1
>>> a[3]
4
>>> a[4]

Traceback (most recent call last):
  File "<pyshell#27>", line 1, in <module>
    a[4]
IndexError: list index out of range
>>> len(a)
4
>>> a[1:3]
[2, 3]
>>> del a[0]
>>> a
[2, 3, 4]
>>> a*3
[2, 3, 4, 2, 3, 4, 2, 3, 4]
>>> a+b
[2, 3, 4, 3, 6, 9, 12]
>>> a = [1,4,7,9]
>>> sum = 0
>>> for i in a:
        sum+=i


>>> sum
21
>>>
```

**CODE 3.6**
List example.

If you try to access an element with an index that is larger than the number of the total elements, then you will get an *IndexError*, indicating the index is out of range (Code 3.6b).

- len[list]: gets the length of the elements (calculates the total number of the element) (Code 3.6c).

```
>>> a = [1,2,3,4]
>>> a[0]
1
>>> a[3]
4
```

**CODE 3.6a**
A List operation.

```
>>> a = [1,2,3,4]
>>> a[4]

Traceback (most recent call last):
  File "<pyshell#27>", line 1, in <module>
    a[4]
IndexError: list index out of range
```

**CODE 3.6b**
List operation out of range.

```
>>> a = [1,2,3,4]
>>> len(a)
4
```

**CODE 3.6c**
List length.

- seq[index1: index2]: generates a new sequence of data type with elements sequenced from index1 to index2 (Code 3.6d).
- *del* seq[index]: deletes the elements sequenced as *index* (Code 3.6e).
- seq * expr (Code 3.6f)
- seq1 + seq2: unions two sequence objects (Code 3.6g)

```
>>> a = [1,2,3,4]
>>> a[1:3]
[2, 3]
```

**CODE 3.6d**
Subset a List.

```
>>> a = [1,2,3,4]
>>> a
[1, 2, 3, 4]
>>> del a[0]
>>> a
[2, 3, 4]
```

**CODE 3.6e**
Delete an element from a List.

```
>>> a = [1,2,3,4]
>>> a*3
[1, 2, 3, 4, 1, 2, 3, 4, 1, 2, 3, 4]
```

**CODE 3.6f**
List multiplies an integer.

```
>>> a = [1,2,3,4]
>>> a = [11,12,13,14]
>>> a + b
[1, 2, 3, 4, 11, 12, 13, 14]
```

**CODE 3.6g**
Union two sequence objects.

```
>>> a = [1,4,7,9]
>>> sum = 0
>>> for i in a:
        sum+=i

>>> sum
21
```

**CODE 3.6h**
Loop each object in the complex data.

- obj in seq (obj not in seq): loops through each object in the complex data and performs an operation with each element. The example goes through each object in the list a, and adds the value of each object to the sum obj (Code 3.6h).

String data type also belongs to the sequence data type, and those operators can be applied to a string object.

**Methods:** As seen in the classes created in the previous chapters, a list is a system built-in class. The objects created from list have many methods. The most important methods include *append(), insert(), remove(), pop(), sort(),* and *reverse()* (Code 3.7).

```
>>> a = [1,2,3,4]
>>> a.append(10)
>>> a                    Append item 10 at the last position
[1, 2, 3, 4, 10]
>>> a.insert(2,15)
>>> a                    Insert item 15 at the index position
[1, 2, 15, 3, 4, 10]
>>> a.pop()
10                       Popup the last item
>>> a
[1, 2, 15, 3, 4]
>>> a.sort()
>>> a                    Sort the list
[1, 2, 3, 4, 15]
>>>
```

**CODE 3.7**
List methods.

**Built-in functions:** There are three common built-in functions for handling a list, which are handy in GIS programming (Khalid 2016):

- *filter*(func, list): Filter is used to extract a target list from the original list. For example, you can use the filter function to select all cities within Virginia, or select the restaurants and hotels within Fairfax city.
- *map*(func, list): Map is used to convert the original list to a new list using the function. For example, you can convert it from degrees to meters.
- *reduce*(func, list): Reduce is another method that is useful for real-world GIS problems. To calculate the total street or road length of Fairfax County, *reduce* can invoke a function *func* iteratively over each element of the list, returning a single cumulative value.
- Tuple

  Similar to lists, *tuple* is another complex data type. One obvious difference between tuple and list is that it is denoted by the use of parentheses. Another difference is that tuple data type is immutable (Table 3.4), meaning that the element cannot be altered once it is defined. An error will occur if the value of a *tuple* element is altered (Code 3.8).
- Dictionary

  A dictionary is mutable and a container data type that can store any Python objects, including other container types. A dictionary differs from sequence type containers (lists and tuples) in how the data are stored and accessed.

```
>>> a = [1,2,3,4]
>>> a
[1, 2, 3, 4]
>>> a[0]=5
>>> a
[5, 2, 3, 4]
>>> x = (1,2,3,4)
>>> x
(1, 2, 3, 4)
>>> x[0]=5          Immutable, cannot be changed

Traceback (most recent call last):
  File "<pyshell#55>", line 1, in <module>
    x[0]=5
TypeError: 'tuple' object does not support item assignment
>>>
```

**CODE 3.8**
Tuple operation.

```
>>> parkingLots = {'A': [-101.12, 32.13, 'parking lot A',
'General parking'], 'B': [-101.14, 32.56, 'parking lot B',
'Stuff only'] }
>>> parkingLots
{'A': [-101.12, 32.13, 'parking lot A', 'General
parking'], 'B': [-101.14, 32.56, 'parking lot B',
'Stuff only']}
>>> parkingLots['A']
[-101.12, 32.13, 'parking lot A', 'General parking']
>>>
```

**CODE 3.9**
Dictionary operation.

**Define:** The syntax used to define a dictionary entry is {key:value, key:value, …..}, where all elements are enclosed in braces. It is convenient for storing spatial feature objects so each object will include a unique object ID, spatial attributes (coordinates), and nonspatial attributes (e.g., Name). Where unique IDs can be used as key, all attributes can be used as a value. The keys are integers or strings while values can be any data type, such as *list* or *dictionary* (Code 3.9).

In Code 3.9, *parkingLots* is declared as a dictionary data type that includes two elements. The unique ID and parking lot sequence *'A'* and *'B'* are used as the keys, and the attribute information of each key is held in a distinct list.

- Set

  *Set* is used to construct and manipulate unsorted collections of unique elements. A *set* object can either be created from {*v1, v2, v3,….*} where elements are surrounded by braces, or from a *set(list)* where the argument is a *list* (Code 3.10).

**Operations:** Set supports several operations (Table 3.5), including union (|), intersection (&), difference (-), and symmetric difference (^) (Linuxtopia 2016).

**Methods:** *Set* is a system built-in class. Several important methods are supported by a *set* object, including add(), remove(), and pop(). It also

```
>>> s = {1,3}
>>> s
set([1, 3])
>>> a = [2,4,3,1]
>>> s = set(a)
>>> s
set([1, 2, 3, 4])
>>>
```

**CODE 3.10**
Set operation.

**TABLE 3.5**

Operations between Two *Set* Objects, *s* and *t*

| Operation | Operator | Function | Usage |
|---|---|---|---|
| difference | − | Create a new set with elements in *s* but not in *t* | |
| intersection | & | Create a new set with elements common to *s* and *t* | Get spatial objects with both conditions matched, such as finding the restaurants within Fairfax, as well as with French style |
| symmetric difference | ^ | Create a new set with elements in either *s* or *t* but not both | |
| union | \| | Create a new set with elements in both *s* and *t* | Combine two states' cities to get collection |

```
>>> s = set(['A','B','C','D'])
>>> t = set(['A','B','E','F'])
>>> s-t
set(['C', 'D'])
>>> s&t
set(['A', 'B'])
>>> s^t
set(['C', 'E', 'D', 'F'])
>>> s|t
set(['A', 'C', 'B', 'E', 'D', 'F'])
>>> s.difference(t)
set(['C', 'D'])
```

**CODE 3.11**
Set operations.

supports four methods: difference(), intersection(), symmetric_difference(), and union() (Code 3.11).

## 3.4 Miscellaneous

### 3.4.1 Variables

A variable is a memory space reserved for storing data and referred to by its name. Before use, a variable should be assigned a value. Variables have different types of values. The basic variable value types include byte, short, int, long, text, float, and double, as introduced in Section 3.4.1. Some types of data can be converted using typecast. For example, float("3.14") will convert texts into a floating number 3.14.

```
>>> class Point:
        def __init__(self,x,y):
            self.x = x
            self.y = y
```

float
```
>>> x = float(1) # Initialize variable x as float
>>> y = float(1) # Initialize variable y as float
>>> p1 = Point(x,y) # Initialize variable p1 as float
>>> p1.x
1.0
```
string
```
>>> x = 'x has a dynamic type' # change x to string
>>> print x
x has a dynamic type
```

**FIGURE 3.3**
Dynamic data type.

In Python, variable types are dynamic, with the type only defined when its value is assigned. This means a variable can change to a different data type. For example (Figure 3.3), x = float(1) will assign x as float, but x = 'x has a dynamic type' will change the variable x from a float type to a string type. Here, we use 'p1' as an object variable name.

A name is required for each variable. The variable's name must be a legal identifier, which is a limited combination series of alphabet letters, digits, and underscores. The name must begin with a character or underscore, but it may not start with a digit. Therefore, 'point1' is a legal name, but '1point' is illegal. In addition, blanks are not allowed in variable name. Python reserved words (Table 3.2) cannot be used as variable name.

## 3.4.2 Code Style

There are several guidelines to Python programming style.

**Meaningful names:** Use meaningful names for variables, classes, and packages. If the meaning is not clear, add a comment to identify what the name means.

**Whitespace in expressions and statements:** Always surround operators with a single space on both sides.

**Indentation:** Indented codes are used to organize code, as introduced in Section 3.2.3.

**Comments:** Comments are used for documenting or explaining code. Python supports two kinds of comments:

*# comments*

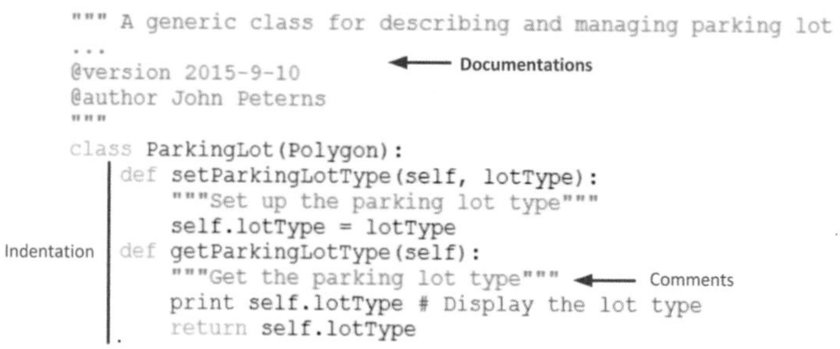

```
""" A generic class for describing and managing parking lot
...
@version 2015-9-10          ◄—— Documentations
@author John Peterns
"""
class ParkingLot(Polygon):
    def setParkingLotType(self, lotType):
        """Set up the parking lot type"""
        self.lotType = lotType
    def getParkingLotType(self):
        """Get the parking lot type"""   ◄—— Comments
        print self.lotType # Display the lot type
        return self.lotType
```

Indentation (label pointing to the indented block)

**FIGURE 3.4**
Coding style.

The compiler ignores everything from the # to the end of the line.

> *""" comments line 1*
> *comments line 2*
> *"""*

This comment style indicates documentation strings (a.k.a. "docstrings"). A docstring is the first statement in a module, function, class, or method definition. All modules should have docstrings, and all functions and classes exported by a module should have docstrings (Figure 3.4).

## 3.5 Operators

Operators include basic characters, division and type conversion, modulo, negation, augmented assignment, and Boolean operations. These operators are categorized into several types:

- Arithmetic Operators (Table 3.6)
- Bitwise&Shift Operators (Table 3.7): Bitwise&Shift Operators can be very complex. A general idea about these operators should be understood and this book will use binary data representation.
- Assignment Operators (Table 3.8): Assignment operators are used to assign values to variables. The most common assignment uses "=" symbol. An example is a = b, which means assign value b to variable a. Assignments can integrate arithmetic and Bitwise&Shift operators yielding complex assignments such as "+=" (Add AND) (Code 3.12).

**TABLE 3.6**

Arithmetic Operators (Assume Variable *a* Holds 5 and Variable *b* Holds 2)

| Arithmetic Operators | Description | Example |
|---|---|---|
| + | Addition | >>> a + b<br>7 |
| − | Subtraction | >>> a − b<br>3 |
| * | Multiplication | >>> a * b<br>10 |
| / | Division | >>> a/b<br>2.5 |
| ** | Exponentiation: Performs exponential (power) calculation on operators | >>> a ** b<br>25 |
| % | Modulus: Divides left-hand operand by right-hand operand and returns remainder | >>> a % b<br>1 |
| // | Floor Division: The division of operands where the result is the quotient in which the digits after the decimal point are removed | >>> a // b<br>2<br>>>> 5.0 // 2.0<br>2.0 |

- Comparison Operators (Table 3.9): Comparison operators are used to compare two variables or objects, to check whether they are equal or different.
- Logic Operators (Table 3.10): Logic operators are used together with *if, else, while* keywords to create logic control statements (Section 3.4). Statements with logic operators are either True or False. Code 3.13 shows an example.

**TABLE 3.7**

Bitwise&Shift Operators (Assume Variable *a* Holds 5 and Variable *b* Holds 2)

| Item | Description | Example |
|---|---|---|
| >> | Binary Right Shift Operator. The left operand's value is moved right by the number of bits specified by the right operand. | a >> b will give 1, which is 0000 0001 |
| << | Binary Left Shift Operator. The left operand's value is moved left by the number of bits specified by the right operand. | a << b will give 20, which is 0001 0100 |
| & | Binary AND Operator copies a bit to the result if it exists in both operands. | a & b will give 0, which is 0000 0000 |
| \| | Binary OR Operator copies a bit if it exists in either operand. | a \| b will give 7, which is 0000 0111 |
| ^ | Binary XOR Operator copies the bit if it is set in one operand but not both. | a ^ b will give 7, which is 0000 0111 |

**TABLE 3.8**

Assignment Operators (Assume Variable *a* Holds 5 and Variable *b* Holds 2)

| Assignment Operators | Name | How to Use | Equivalent | Result |
|---|---|---|---|---|
| = | Equal assignment | a = b | Set a as b | 2 |
| += | Add AND | a+ = b | a = a+b | 7 |
| −= | Subtract AND | a− = b | a = a − b | 3 |
| *= | Multiply AND | a* = b | a = a * b | 10 |
| /= | Divide AND | a/ = b | a = a/b | 2 |
| %= | Modulus AND | a % = b | a = a % b | 1 |
| ** = | Exponent AND | a ** = b | a = a ** b | 25 |
| \|= | Binary OR AND | a\| = b | a = a \| b | 7 |
| ^= | Binary XOR AND | a^ = b | a = a ^ b | 7 |
| <<= | Left shift AND | a<< = b | a = a << b | 20 |
| >>= | Right shift AND | a>> = b | a = a>>b | 1 |

```
>>> b=3
>>> a=b
>>> a
3
>>> a+=b
>>> a
6
```

**CODE 3.12**
Add AND example.

**TABLE 3.9**

Comparison Operators (Assume Variable *a* Holds 5 and Variable *b* Holds 2)

| Comparison Operators | How to Use | Compare (or Check) | Results |
|---|---|---|---|
| == | a== b | a equals b | (a == b) is not true. |
| < | a < b | a is less than b | (a < b) is not true. |
| > | a > b | a is greater than b | (a > b) is true. |
| >= | a >= b | a is greater than or equal to b | (a >= b) is true. |
| <= | a <= b | a is less than or equal to b | (a <= b) is not true. |
| != | a != b | a is not equal to b | (a != b) is true. |
| is | a is b | a and b are the same object | (a is b) is not true. |
| is not | a is not b | a and b are different objects | (a is not b) is true. |
| in | a in range(b) | a is a member of [1, 2, …,b] | (a in range (b)) is not true. |
| not in | a not in range (b) | a is not a member of [1, 2, …,b] | (a not in range (b)) is true. |

**TABLE 3.10**

Logic Operators (Assume Variable *a* Holds True and Variable *b* Holds True)

| Logic Operators | How to Use | Results |
|---|---|---|
| And | Logical AND operator. If both the operands are true, then condition becomes true. | (a and b) is true. |
| Or | Logical OR operator. If any of the two operands are nonzero, then condition becomes true. | (a or b) is true. |
| Not | Logical NOT operator. Use to reverse the logical state of its operand. If a condition is true, then Logical NOT operator will make false. | Not (a and b) is false. |

```
x = int(raw_input('Enter your input for x: '))
y = int(raw_input('Enter your input for y: '))
if x >5 and y > 5:
        print 'both x and y are bigger than 5'
if x > 5 or y > 5:
        print 'either x or y is bigger than 5'
if   x > y:
        print 'x is bigger than y'
if   not   x > y:
        print 'x is smaller than y'
```

**CODE 3.13**
Logic operations.

## 3.6 Statements

A statement is a combination of variables and operators. The statement should comply with the operator's usage. If you assign a value, you should use assignment operator. If you accidentally use comparison operators, you should expect an error. Pay attention to the statement's precision. For example (Code 3.14), the first and second i seem to be assigned with similar values using identical division and addition operations. However, they generate different results.

```
>>> i = 1/2+1/2
>>> i
0
>>> i = 1.0/2 + 1.0/2
>>> i
1.0
>>>
```

**CODE 3.14**
Statement examples.

## 3.7 Functions

Functions are defined by the keyword *def* followed by the function name with various input arguments. Similar to the methods defined within a class, the number of arguments can be zero to many. Class methods are special functions with the first argument as 'self.'

  def funcName(arg1, arg2, …):

  Statement blocks

A function is defined by a function name and other required input arguments, like funcName(arg1, arg2, …). The arguments passed to a function while calling a function should match what is declared. A function can return none, one, or more arguments (Code 3.15).

When you declare default arguments for class methods (Section 2.1), you can also set up default values for function arguments, and make those arguments optional for the caller. The example (Code 3.16) shows how to calculate cost function (calCost) and taxRate as the default argument. Therefore, when the calCost function is set with the parameter as 100, the variable taxRate uses 0.05 as its value.

*Tips*

Use the keyword lambda to declare one line version of a function. Usually such functions are anonymous because they are not defined in a standard manner. The body of the lambda function statement should be given on the same line, like in the add() function (Code 3.17).

```
>>> # Return nothing
>>> def hello():
        print 'Hello, World!'


>>> hi = hello()
Hello, World!
>>> print hi
None
>>> #Return a value
>>> def add(x,y):
        return x+y


>>> z = add(1,2)
>>> print z
3
>>>
```

CODE 3.15
Return value from a function.

```
>>> #Default Arguments
>>> def calCost(price, taxRate = 0.05):
        return price + price*taxRate

>>> calCost(100)
105.0
>>> calCost(100,0.075)
107.5
>>>
```

**CODE 3.16**
Default arguments.

```
>>> def add(x,y):
        return x+y

>>> add(1,2)
3
>>> a = lambda x,y:x+y
>>> b = a(1,2)
>>> print b
3
```

**CODE 3.17**
Lambda example.

The Python interpreter has built-in functions that are always available. They are listed in alphabetical order in Figure 3.5. The functions written in red have already been introduced. The functions written in blue are important, and will be introduced in later chapters.

| abs() | divmod() | input() | open() | staticmethod() |
|---|---|---|---|---|
| all() | enumerate() | int() | ord() | str() |
| any() | eval() | isinstance() | pow() | sum() |
| basestring() | execfile() | issubclass() | print() | super() |
| bin() | file() | iter() | property() | tuple() |
| bool() | filter() | len() | range() | type() |
| bytearray() | float() | list() | raw_input() | unichr() |
| callable() | format() | locals() | reduce() | unicode() |
| chr() | frozenset() | long() | reload() | vars() |
| classmethod() | getattr() | map() | repr() | xrange() |
| cmp() | globals() | max() | reversed() | zip() |
| compile() | hasattr() | memoryview() | round() | __import__() |
| complex() | hash() | min() | set() | apply() |
| delattr() | help() | next() | setattr() | buffer() |
| dict() | hex() | object() | slice() | coerce() |
| dir() | id() | oct() | sorted() | intern() |

**FIGURE 3.5**
System built-in functions. (From Python. 2001a. Built-In Functions. https://docs.python.org/3/library/index.html (accessed September 3, 2016).)

```
>>> numbers = range(1,101,1)
>>> sum1, sum2, sum3 = 0, 0, 0
>>> for i in numbers:
        sum1=sum1+i


>>> print sum1
5050
>>> i = 0
>>> while (i<101):
        if (i%2==0):
                sum2=sum2+i
        i+=1


>>> print sum2
2550
>>> for i in range(100):
        if (i%2==1):
                sum3+=i


>>> print sum3
2500
>>>
```

**CODE 3.18**
Sum-up calculations.

## 3.8 Hands-On Experience with Python

The Python program (Code 3.18) calculates the sum of 1–100, all odd numbers between 1 and 100, and all even numbers between 1 and 100. Can you solve for each of those calculations? Try coding in Python GUI and explain how each calculation works and the structure of the codes.

## 3.9 Chapter Summary

This chapter introduces

- Python syntax
- Python data types
- Operators
- What a function is and how to declare a function

**PROBLEMS**

1. Keywords: Check the following keywords (Table 3.2.) and briefly explain them (concepts, when/how to use keywords [use function help() to get help about each keyword, e.g., help('if')], and design/program an example of how to use the keywords). For example, if, elif, else keywords.

   a. Explain

      if, elif, else are the keywords used to make decisions......

   b. Examples (Code 3.19)

2. Operators: The five categories of operators include arithmetic operators, shift & bitwise operators, assignment operators, comparison operators, and logic operators. For each pair indicated below, explain the differences between the two operators and then design and enter an example in Python interpreter to demonstrate the difference.

   a. "+" vs. "+="

   b. "%" vs. "/"

   c. "*" vs. "**"

   d. "==" vs. "is"

   e. "!=" vs. "is not"

   f. " in " vs. "not in"

   g. " and " vs. "or"

   h. " not in " vs. " not"

3. Class and Object Concepts

   a. Briefly describe the argument "self" in class method and provide an example.

```
>>> x = [1,2,3,4,5]
>>> y = 8
>>> z = [6,7,8,9]
>>> if y in x:
        print y, 'is in', x
elif y in z:
        print y, 'is in z', z
else:
        print y, 'is not in either x or z', x, z


8 is in z [6, 7, 8, 9]
>>>
```

**CODE 3.19**
If...elif...else example.

b. Design a simple class including at least one static, one private, one protected, and one public attribute (which could also be a method).

c. Create an object and then access and print those attributes.

4. Data Types and Functions

In Python, the seven common data types are as follows: integer, long integer, float, string, list, tuple, and dictionary. The three basic types of vector data are Point, Polyline, and Polygon. Try to implement the following class design of Polygon in Python using the proper Python data types (use other data types instead of float) for the attributes and methods.

Description: A Point class has two attributes, x and y, to represent the coordinate of a point. A Point class also has a method calDis () to calculate the distance between two points. The arguments for the calDis method is point object 'self,' and the another point object is "point." The return value for the distance between two points is designed as float.

The UML design for Point class to keep point vector data is as follows:

| Point(Feature) |
| --- |
| +x: float |
| +y: float |
| + calDis(p: Point): float |

The following Code 3.20 exemplifies how to implement the Point class:

Description: A Polygon class has one attribute "points" to represent the list of coordinates. A Polygon class also has a method getLength () to calculate the perimeter of the Polygon. The arguments for the getLength method is current Polygon object 'self.' The return value for the border length of a polygon is designed as float.

```
>>> import math
>>> class Point:
        def __init__(self, x = 0, y = 0):
                self.x = x
                self.y = y
        def calDis(self,point):

                return math.sqrt((self.x-point.x)**2+(self.y-point.y)**2)

>>>
```

CODE 3.20
Point class definition.

The UML design for a Polygon class keeping polygon vector data is as follows:

| Polygon(Feature) |
| --- |
| +points: list<Points> |
| + getLength(): float |

Assign a polygon with the following data: [(1.0, 2.0), (3.0, 5.0), (5.0, 6.0), (1.0, 2.0)] and calculate the border length of the polygon.

# 4

# *Python Language Control Structure, File Input/Output, and Exception Handling*

In general, computer program statements are executed sequentially. Control structures help computer programs break the sequence by jumping back and forth to skip a set of statements. The control structures make programming languages much more versatile for solving real-world problems. Control structures include making decisions and loops. This chapter introduces frequently used Python control structures and demonstrates methods for operating GIS objects. This chapter also helps you learn how to read data from and/or write data to a local file. When a Python program encounters an error event (a.k.a. exception), it will crash if not handled. This chapter briefly covers how to capture and handle exceptions.

## 4.1 Making Decisions

Decisions are one of the most basic ways to change program execution flow. They use conditional expressions to evaluate whether a statement is True or False. For example, assuming a = 3 and b = 2, the conditional expression a > b is True, while the expression a < b is False.

There are several expressions/values that result in False:

- Any number with a value of zero (e.g., 0, 0.0, 0L, 0j, Code 4.1 right)
- An empty string (" or "")
- An empty container, such as list (Code 4.1 left), tuple, set, and dictionary
- False and None

Conditional expressions consist of operators and relevant variables/values. Comparison and logic operators are used to construct conditional expressions and statements.

- *Comparison Operators*: >, >=, <, <=, = =, !=, is, is not, in, not in
- *Logic Operators*: and, or, not

```
>>> if []:                          >>> if 0:
        print True                          print True
else:                               else:
        print False                         print False

False                               False

>>>
```

**CODE 4.1**
False conditional expressions: empty list (left) and zero (right).

Logic operators are used to provide multiple conditional expressions, for example, if x>0 and y>0: print 'x and y are positive', or perform negation of expressions using not.

- *if statement*: When one condition is used to control the statement execution, an *if statement* is used to execute the statement block based on the evaluation results of the condition expression. When a condition expression is True, the following statement block will be executed; otherwise, the following statement block will be skipped.

Statement syntax

if (conditional expression):
    Statement block

Statement example:

if a > b:
    print "a is bigger than b"

- *if…else statement*: When two conditions result in different executions, the *if statement* is used with the *else* clause. If the conditional expression is True, then the statement block following *if* will be executed; otherwise, the statement block after *else* will be executed.

Statement syntax:

if (conditional expression):
    Statement block
else:
    Statement block

Statement example:

if a > b:
    print "a is bigger than b"

else:
    print "a is smaller than b"

- *if….elif…else statement*: When more than two conditions result in different executions respectively, use *if, elif,* or *else*. The *elif* statement is similar to the *else if* used in C or Java. This statement will allow programmers to check on multiple conditions.

elif syntax:

if (conditional expression 1):
    Statement block 1
elif (conditional expression 2):
    Statement block 2
elif (conditional expression 3):
    Statement block 3
…
else:
    Statement block n

If the conditional expression 1 (or 2, 3,….) is true, then the statement block 1 (or 2, 3….) will be executed and the other statement block will be skipped. However, if all above conditions (1, 2, 3, …., n–1) are not true, the blocks under *else* (statement block n) will be executed.

*Tips*: pass statement (Code 4.2)

*pass* statement is unique in that it does not perform any function. It is used in the decision-making process, telling the interpreter not to do anything under certain conditions.

In the software development process, it can serve as a place holder, to be replaced later with written code (Code 4.3).

```
a=b=0
if a>b:
    pass
else:
    pass
```

**CODE 4.2**
Pass statement in if … else… structure.

```
def draw():
    pass
```

**CODE 4.3**
Pass is used as a place-holder statement written in method.

## 4.2 Loops

Another type of control structure is the loop. Usually, a loop executes a block until its condition becomes false or until it has used up all the sequence elements in a container (e.g., list). You can either interrupt a loop to start a new iteration (using continue) or end the loop (using *break*). Both *while* and *for* can be used to loop through a block of statements.

- *for statement*: *For* loop is used to execute a repeated block of statements for a definite number of times. *For* is used with composite data types (also known as sequence data types), such as string, list, and tuple in the following syntax and example, which uses a for loop to calculate the sum of 1 to 100 (Code 4.4):

*Syntax:*

for item in sequence:
    Statement block

- *while statement*: *while* statement is very flexible, and can repeat a block of code while the condition is true. The *while* statement is usually applied when there is an unknown number of times before executing the loop. Code 4.5 shows an example of using *while* loop to calculate the sum of 1 to 100.
- *range() function and len() function*: The *range* (Pythoncentral 2011) and *len* functions are often used in *for* and *while* loops. Using *range*(start, end, step) generates a list where for any k, start <= k < end, and k iterates from start to end with increments of step. For example, range(0,4,1) produces a list of [0,1,2,3]; range(0,50,10) produces a list of [0,10,20,30,40]. *range* function takes 0 as default starting value and 1 as default step. For example, range(4) produces a list of [0,1,2,3]. Code 4.4 is an example using range(4) to produce a list, and using for loop structure to print every element within the list.

```
>>> for i in range(4):
        print i


0
1
2
3
```

**CODE 4.4**
Use range function with default start and step values.

```
>>> i=0
>>> total=0
>>> while i<101:
            total +=i
            i+=1


>>> print total
5050
>>>
```

**CODE 4.5**
Calculating summary of 1 to 100 using while loop.


The following example illustrates how to use the range function, and how to calculate the sum of 1 to 100 by using the *total* variable to hold the summarizing result for loop (Code 4.6).

The function *len()* returns the total number of elements in composite data. For instance, len(polyline.points) can return the number of points within a polyline. The following example uses *while* and *len()* to calculate the length of a polyline (Code 4.7).

- *break*: A *break* is used to interrupt the execution of a loop, such as finding if two lines intersect one another. The loop can be broken once the two line segments (from each of the two lines) intersect (Code 4.8).

```
>>> total = 0
>>> for i in range(1,101,1):
            total+=i


>>> print total
5050
```

**CODE 4.6**
Calculate the summary of 1 to 100 using range and for loop.

```
>>> def getLength(polyline):
            length, i=0.0, 0
            while i<len(polyline.points)-1:
                    length+=math.sqrt((polyline.points[i].x-
                    polyline.points[i+1].x)**2 +
                    (polyline.points[i].y-polyline.points[i+1].y)**2)
```

**CODE 4.7**
Calculate the length of a polyline using *while* loop and *len()* method.

```
>>> bIntersect = False
>>> def intersect(line1,line2):
        for i in len(line1.lineSegments):
            for j in len(line2.lineSegments):
                if (line1.lineSegments[i].segIntersect
                    (line2.lineSegments[j])):
                    bIntersect = True
                    break
```

**CODE 4.8**
Test if two lines intersect with each other using *break* and *for* loop.

```
>>> def drawMap():
        for i in range(len(layers)):
            if (layers[i].layerType == 'Image'):
                continue
            layers[i].drawLayer()
```

**CODE 4.9**
Draw a map without drawing the image layers.

- *continue*: The *continue* statement is used less often than the break statement to skip the rest of a loop body under a certain condition. This can eliminate executions of specific loop values or categories of values. For example, when a map is drawn, you may uncheck the layers with an image. The code (Code 4.9) skips all image layers:

## 4.3  Other Control Structures

- *Loop and decisions combination*: Loop and decisions are used together to construct complex flow control:

for/while conditional expression 1:

  statement blocks 1
  if conditional expression 2
      statement blocks 2
  else:
      break/continue
      statement blocks 3

The following example (Code 4.10) identifies the points shared between two polygons (this is helpful to understand the data structure, although tuple is not normally used to hold Point coordinates):

```
>>> p1, p2, p3, p4 = (0,0), (1, 1),(2, 2), (3, 3)
>>> polygon1= [p1, p2, p3]
>>> polygon2 = [p1, p2, p4]
>>> def getCommonList(list1, list2):
        commonList = []
        for eachVal in list1:
            if eachVal in list2:
                commonList.append(eachVal)
        return commonList

>>> results = getCommonList(polygon1, polygon2)
>>> print results
[(0, 0), (1, 1)]
```

**CODE 4.10**
Loop and decisions combination example.

```
>>> def doubleloop():
        for i in range(len(points)-1):
            for j in range(len(points)-i-1):
                points[i].calDis(points[len(points)-1-j])
```

**CODE 4.11**
Calculate distance between points using double loops.

- *Nested loops*: Another complex structure are nested loops, especially double loops. Given a list of points, if you were to calculate the distance of each pair of points on a point list, you will need to use the double loop (Code 4.11).

## 4.4 File Input/Output

There are a series of built-in methods for operating files, such as open, read, and write (PythonForBeginers 2012).

- *Opening a file*: To operate a file, use the function *open* with a filename and file operation mode as arguments. In the example f = open(filename, mode), mode could be
  - 'r': when the file will only be read, and this is the default value.
  - 'w': for only writing (an existing file with the same name will be erased).

- 'a': open the file for appending; any data written to the file are automatically added to the end of file.
- 'r+': open the file for both reading and writing.
- *Reading a file*: The *open()* function will load the file from disk storage to the memory, and return it as a file object. This file object has three file read methods:
  - *read*(size): returns the entire file when no size parameter is passed (by default the size is equal to –1), or content as a string in the byte size specified.
  - *readline*(): reads and returns one line as a string (including trailing '\n').
  - *readlines*(): reads and returns all lines from file as a list of strings (including trailing '\n').
- *Writing a file*: A file object has two written methods:
  - *write*(str): writes string 'str' to file.
  - *writelines*(list): writes a list of strings to file; each string element is one line in the file.

The *write* function writes the contents of a string to a file. However, there are other data types such as float and integer. How is data written into a file? Since a string acts as the input argument, the *str()* and *repr()* function will convert a nonstring object to a string object. For example, if you write a float number 3.14 into a file, you can use *write(str(3.14))*. Typically, a text file is organized line by line, while the *write()* function writes data into a file and changes it to a new line using the special characters "\n". Thus, write ("first line\nsecond line") will output two lines in the file as shown below:

first line

second line

The return value is also a string when reading the data from a file with *read()* and *readline()*; therefore, we need to format those string values into the data type we prefer. We can use *float*(str), for example, *float*('3.14'), *int*(str), for example, *int*('2.0'), and *long*(str), for example, *long*('2.0l') to convert strings to numbers.

- *Change file object's pointer position*: While reading a file, we may need to skip several lines and read out specific information in the file. Under such circumstances, we can locate specific lines and words of a file with the following two methods:

- *seek*(offsize, whence): go to a position within a file, with offsize bytes *offset* from *whence* (0==beginning of file, 1==current location, or 2==end of file).
- *tell*(): return current cursor/pointer position within a file.
- *Close a file*: After we finish manipulating a file, we should close the file to release it from the memory with the method:
  - *close*(): close a file

This is a very important step whenever we read or write data to a file. Without this step, the file will be kept in the memory and may exhaust the available system memory (memory leak) if too many files are opened. Data may not be preserved onto hard drive before closing a file.

## 4.5 Exceptions

Exceptions (Python 2001b) are the errors encountered when executing Python programs, for example, the errors to open a nonexisting file, division by zero, concatenate 'str' and 'int' objects. These exceptions can be handled in Python programs using the *try...except...* statement. There can be one or more *except* clauses in the statement. Each *except* clause is used to catch one exception. If an exception clause matches the exception, the program will execute the *except* clause. If no *except* clause matches, the program will be passed on to outer *try* statements and give the exception error. Code 4.12 handles the ZeroDivisionError using the *try...except...* statement.

```
>>> def slope(x1,y1,x2,y2):
        try:
                return (y2-y1)/(x2-x1)
        except ZeroDivisionError:
                print 'Error: x1 equals x2'
                return None


>>> slope(1,2,3,4)
1
>>> slope(1,4,1,5)
Error: x1 equals x2
>>>
```

**CODE 4.12**
Handle ZeroDivisionError using *try...except...* statement.

## 4.6  Hands-On Experience with Python

### 4.6.1  Find the Longest Distance between Any Two Points

- Open a text file
- Enter the Code 4.13, where a single loop is used
- Try and test another method where multiple loops are used (Code 4.14).

In this method, double loop is used to calculate the distances between each pair of four points (p0, p1, p2, p3). As shown as Figure 4.1, during the outer loop level 1, when i=0, the inside loop will iterate j from 1 to 2 to 3. After the outer loop advances to i=1, j will be iterated from 2 to 3, and so forth.

### 4.6.2  Hands-On Experience: I/O, Create and Read a File

1. Open the Python console. Type the Code 4.15 for windows OS:

```
>>> import math
>>> class Point: ## define a point class
        def __init__(self, x=0.0, y = 0.0):
            self.x = x
            self.y = y
        def getDistance(self,other): ## declare getDistance
        as a method
            return math.sqrt((other.x-self.x)**2+(other.y-self.y)**2)


#Declare three points
>>> p1,p2,p3 = Point(1,5), Point(2,8), Point(10,3)
## calculate the distances among random two points and keep
them in a list
>>> dist1 = p1.getDistance(p2)
>>> dist2 = p1.getDistance(p3)
>>> dist3 = p2.getDistance(p3)
>>> distances = [dist1,dist2,dist3]
##Declare the biggestDistance variable
>>> biggestDistance = 0.0
>>> for i in range(len(distances)):
        currentDistance = distances[i]
        if currentDistance > biggestDistance:
            biggestDistance = currentDistance


## Finish finding and print
>>> print 'biggest distance is ->', biggestDistance
biggest distance is -> 9.43398113206
>>>
```

**CODE 4.13**
Find the longest distance between any two points of 3 points using single loop.

```
>>> import math
>>> class Point: ## define a point class
        def __init__(self, x=0.0, y = 0.0):
            self.x = x
            self.y = y
        def getDistance(self,other): ## declare getDistance as a method
            return math.sqrt((other.x-self.x)**2+(other.y-self.y)**2)


#Declare four points
>>> p0, p1,p2,p3 = Point(), Point(1,5), Point(2,8), Point(10,3)
## keep in the list
>>> points = [p0, p1, p2, p3]
##Declare the biggestDistance variable
>>> biggestDistance = 0.0
>>> for i in range(len(points)):
        for j in range(i+1, len(points)):
            currentDistance = points[i].getDistance(points[j])
            if currentDistance > biggestDistance:
                biggestDistance = currentDistance


## Finish finding and print
>>> print 'biggest distance is ->', biggestDistance
biggest distance is -> 10.4403065089
```

**CODE 4.14**
Calculate the longest distance between any two points of a 4 point set using double *for* loop
with i and j as incremental variables.

> Note: (1) Make sure the directory 'c:/code' exists, or you will gener-
> ate an error such as: "IOError: [Errno 2] No such file or directory:
> 'c:/code/points.txt'". makedirs() function in os module could help to
> create directory; (2) Make sure you have the permission to create a
> file under 'c:/code' directory, otherwise you will generate an error
> such as "IOError: [Errno 13] Permission denied: 'c:/code/points.txt'".

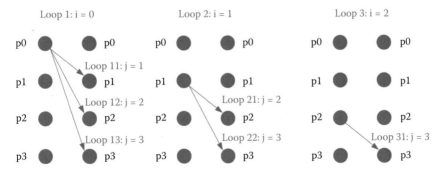

**FIGURE 4.1**
Use double loop to calculate the distance between each pair of four points p0, p1, p2, p3.

```
>>> f = open ('C:/Users/Phil/Downloads/points.txt', 'w+')
>>> f.write('point:\n')
>>> f.write('p1: 1.0, 1.0 \n')
>>> f.write('p2: 2.0, 2.0 \n')
>>> f.close()
```

**CODE 4.15**
Write a text file.

```
>>> f = open ('C:/Users/Phil/Downloads/points.txt', 'r')
>>> f.readline()   # Read the 1st  line
>>> f.readline()  # Read the 2nd line
>>> f.readline()   # Read the 3rd line
>>> f.readline ()  # end of the file
>>> f.seek(0)   #go to the begin of file
>>> f.readline()
>>> f.readlines() # read rest lines in a list
>>> f.seek(0)
>>> f.read() # read rest lines as a string
>>> f.close()
```

**CODE 4.16**
Read from a text file.

For Mac, or Linux OS, replace the directory 'c:/code/points.txt' with a different one, for example, '/home/points.txt'.

2. Go to the 'points.txt' directory, and check that there is a points.txt created (which should be true if there is no error popping up for your code). Now open the file and examine the information in the file.

3. Code 4.16 reads out the data from the points.txt.

## 4.6.3 Hands-On Experience: I/O, Flow Control, and File

- Read out and manipulate point datasets: create a text file under c:/code/points.txt, and put the following three lines in the file:

point:

p1: 1.0, 1.0

p2: 2.0, 2.0

Note: This is a simple GIS data format file, where the first line "point:" indicates that this is a point feature file.

- Read out the points p1 and p2 from the points.txt file and print out the x and y coordinates. Open Python text editor and create a new file to save as read_point.py. Enter Code 4.17 to the read_point.py.

- Question: If we revise the points.txt to the following format (replace the colon with semicolon):

point;

p1; 1.0, 1.0

p2; 2.0, 2.0

```
"""
GGS 650 Lecture 4 Practice
readPointFile() is the function to parse the following format data:
point:
p1: 1.0, 1.0
p2: 2.0, 2.0\n
readPolylineFile() is the function to parse the polyline format as:
polyline;
1: 1.0, 1.0; 2.0, 2.0;....
2: 2.0, 2.0; 3.0, 3.0;....
"""
>>> import math
>>> class Point:   ## define a point class
        def __init__(self, x=0.0, y=0.0):   ## init method for point class
                self.x = x
                self.y = y
        ## Declare getDistance as method of Point
        def getDistance (self, other):
                return math.sqrt((other.x - self.x)**2 +
                (other.y - self.y)**2)
>>> def readPointFile(fileName):
        file = open(fileName,'r')
        #declare empty list for keeping points, and index for line Num
        points,index = [],0
        for line in file: ## Read each line iteratively
            index += 1 ## Increase index after reading one line
            if index == 1:
                continue ## "Ignore the first line 'point\n'"
            # split the line and get the coordinate,e.g,1.0, 1.0
            coords = line.split(':')[1]
            ## Get the point x, y value
            xCoord = coords.split(',')[0]
            yCoord = coords.split(',')[1]
            point = Point(float(xCoord),float(yCoord))
            points.append(point)
        file.close() # remember to close file after reading
        return points
## Call the function for parsing the point file
>>> points = readPointFile('points.txt') #get all points
#print points
>>> length = len(points) # get the length of points list
>>> for i in range(length):
        point = points[i]
        print point.x, point.y ##print the x, y value of each point
1.0 1.0
2.0 2.0
10.0 11.0
11.2 13.4
```

**CODE 4.17**
Read a formatted GIS point data file.

Which portion of the code should we revise? How about the following
format?

point:
1.0, 1.0; 2.0, 2.0

### 4.6.4 Hands-On Experience: Input GIS Point Data from Text File

Code 4.18 reads point text file and parse the data by generating point objects. The maximum distance between any two points is then found and written to a result file. Please try it on your computer and interpret the example line by line.

```
>>> import math
>>> class Point:
        def __init__(self,x,y):
                self.x = x
                self.y = y
        def dis(self,point):
                return math.sqrt((self.x-point.x)**2+
                (self.y-point.y)**2)
>>> f = open('points.txt','r')
>>> f.readline()
>>> i = 0
>>> points = []
>>> while (i==0):
        line = f.readline()
        if (line.find(':')!=-1):
                cords = line.split(':')[1]
                if (cords.fiand(',')!=-1):
                        xy = cords.split(',')
                        points.append(Point(float(xy[0]),
                        float(xy[1])))
        else:
                i=1
>>> outf = open('pointsResults4.17.txt','w')
>>> lpoints = []
>>> dis = 0
>>> print points
[<__main__.Point instance at 0x03373198>,
 <__main__.Point instance at 0x03373990>,
 <__main__.Point instance at 0x03373148>,
 <__main__.Point instance at 0x03373B98>]
>>> for k in range(len(points)):
        for l in range(k+1, len(points)):
                if (points[k].dis(points[l])>dis):
                        dis = points[k].dis(points[l])
                        while (len(lpoints)>0):
                                lpoints.remove(lpoints[0])
                        lpoints.append(points[k])
                        lpoints.append(points[l])
>>> outf.write('The longest distance is between point
['+str(lpoints[0].x)+','+str(lpoints[0].y)+'] and ' +
            'point ['+str(lpoints[1].x)+','+str(lpoints[1].y)+']\n
The distance is '+str(dis))
>>> outf.close()
>>> f.close()
```

CODE 4.18
Read and write the content of a point data file.

Text file content:

Point:

1: 1, 2

2: 100, 300

3: 4, 5

4: 0, 500

5: 10, 400

6: 600, 20

7: 500, 400

8: 500, 500

## 4.7 Chapter Summary

This chapter introduces different flow control structures including

- Make decisions using *if, else,* and *elif.*
- Loop a block of statements using *while* and *for* flow control.
- Combine loop (*for* and *while*) and conditional statements (*if, else,* and *elif*) for complex flow controls.
- Read and write text files.
- Catch and handle errors using *try…except…* statement.

**PROBLEMS**

1. Review the Python tutorial "Input and Output," which came with Chapter 7 of the Python software help document.
2. Analyze the patterns of the following text string and save it to a text file, for example, polylines.txt.

**Polyline:**

1. 1603714.835939442,142625.48838266544; 1603749.4678153452,142620.21 243656706; 1603780.3769339535,142607.37201781105; 1603801.47584667 8,142582.27024446055; 1603830.4767344964,142536.14692804776;

2. 1602514.2066492266,142330.66992144473; 1602521.4127475217,142414.9 2978276964; 1602520.1146955898,142433.93817959353; 1602501.3840010 355,142439.12358761206; 1602371.6780588734,142417.84858870413; 1602

351.6610373354,142408.02716448065;  1602334.5180692307,142388.58748
627454;  1602331.6999511716,142376.66073128115;  1602334.8067251327,
142348.965322732;  1602338.308919772,142323.6111663878;  1602349.022
6452332,142314.50124930218;  1602363.9090971674,142310.79584660195;
1602514.2066492266,142330.66992144473;

3. Write a Python program to parse the text file and use list to hold the two polylines. Please refer to Section 5.6.1 in Python Library Reference (from Python help document) String methods for split(), strip(), and Built-in Functions float(x).

4. Generate two polyline objects.

5. Calculate the length of the two polylines.

6. Review the class materials on handling exceptions and Python tutorial "Errors and Exceptions" (Section 8.3 in Python help document).

7. While reading the file and converting the string data to another data type, for example, float, please add *"try...except...finally..."* to catch the Exceptions, for example, IOError and ValueError.

# 5

## Programming Thinking and Vector Data Visualization

Programming thinking is critical to mastering a program language (Eckerdal et al. 2005). The only way to become accustomed to thinking in a programming capacity is through experience. This chapter will visualize vector data to illustrate how to think in a programming manner.

### 5.1 Problem: Visualizing GIS Data

Visualizing GIS data is like drawing or painting on a blank cloth or canvas (Hearnshaw and Unwin 1994). GIS maps include essential elements such as a legend, map title, pertinent information, and an image of visualized data. An image of a map will include different types of GIS data, used in different ways (Figure 5.1, for example).

Patterns are referred to as symbols, and include geometric patterns (squares, lines, polygon), colors (green, gray, and beige), and a name of the data features. Depending on the amount of computer memory, visualizing GIS data in map form is a tedious process because the data must be put in sequential order. Usually, the drawing process begins with a polygon to a polyline, and then to a point feature. This way, the smaller areal features will not cover any big areal features. Map drawing processes include multiple steps, as denoted in Figure 5.2.

Prepare Canvas → prepare data/feature to draw (coordinate transfer) and know where to draw → setup how to draw and draw the features → finalize a drawing process (drying, etc.)

A computer program will require a GUI to interact with an end user for input and output visualization. For example, the Python IDLE window is a type of GUI used to program in Python. ArcGIS has both an ArcMap and other GUIs for user interaction. In GIS software, both the GUI and GIS legend are important in interpreting functions needed for a GIS system. This section will discuss how to set up and use a simple GUI user interaction, as well as prepare Canvas for map drawing.

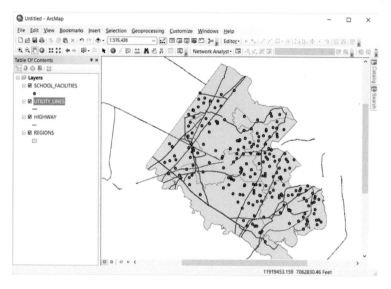

**FIGURE 5.1**
A map using ArcGIS software, in which polygons are illustrated in light green, polylines in dark red, and points in green.

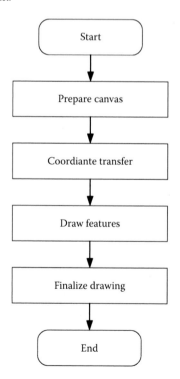

**FIGURE 5.2**
The process for visualizing GIS data.

In Python, Tkinter and related classes and modules are commonly used for drawing and developing GUIs. Aside from the Tkinter, there are other toolkits available, like PyGTK, PyQt, and wxPython. Code 5.1 will bring up a GUI:

In the sample code:

1. The first line will import Tkinter's classes and functions.
2. The second line calls Tk() to generate a root window object.
3. The third line calls Label class to generate a widget Label object.
4. The fourth line uses pack() to make the label visible.
5. The last line brings up the first window, which includes the label (done when label is created).

In this example, a window was created with one label showing "Hello World." TKinter supports a variety of widget objects; the most common are described in Table 5.1.

In addition to the widgets in Table 5.1, Tkinter has other widgets, which include Entry, Frame, LabelFrame, Menubutton, OptionMenu, panelWindow, Scale, Spinbox, and Toplevel.

When using other widgets, replace the third line and fourth line (in Code 5.1) by creating an object of a specific widget and passing it in relevant arguments specified on the online reference,* for example, when using Canvas for map, replace the third line and fourth line with

can = Canvas(root, width = 800, height = 600)

can.pack()

Among all the widgets, Canvas is the most widely used. It supports many methods, like drawing points, lines, polylines, and polygons. A typical Canvas preparing code is shown in Code 5.2.

In Code 5.2, the third line creates a Canvas with the pixel size dimensions of 800 by 600. Although the size is based on the computer monitor size, the actual size can be adjusted on the display settings tab. The second to last line ensures that Canvas is visible on the window and the last line

```
>>> from Tkinter import *
>>> root = Tk()
>>> w = Label (root, text="Hello, world!")
>>> w.pack()
>>> root.mainloop()
```

**CODE 5.1**
Create a GUI using Tkinter.

---

* http://www.pythonware.com/library/tkinter/introduction/tkinter-reference.htm

**TABLE 5.1**

Popular TKinter Widgets

| Widgets | Syntax | Example |
|---------|--------|---------|
| Button | Using (master, options) as parameters to initialize a widget, root is the parent widget, options are the widget options such as command, back | Button(root, text="OK", command=callback) |
| Canvas | | Canvas(root, width = 800, height = 600) |
| Label | | Label(root, text="Hello, world!") |
| Listbox | | Listbox(root, height=8) |
| Menu | | Menu(root, borderwidth=2) |
| Message | | Message(root, text="this is a message") |
| Radiobutton | | Radiobutton(root, text="Grayscale", value="L") |
| Scrollbar | | Scrollbar = Scrollbar(root) |
| Text | | Text(root, font=("Helvetica", 16)) |

```
>>> from Tkinter import *
>>> root = Tk()
>>> can = Canvas(root, width=800, height = 600)
'''draw features and maps on the Canvas/cloth
...
...
'''
>>> can.pack()
>>> root.mainloop()
```

**CODE 5.2**
Preparing a Canvas for drawing.

will ensure the window shows up. The Canvas prepared can be used as the map cloth, which can be used to draw points, polylines, and polygons. The next step is to prepare GIS data so that it can serve as visualization on Canvas.

## 5.2 Transforming Coordinate System

GIS data are represented in a variety of formats and projections, for example, a shapefile in ArcView versus a text file formatted in different applications. The first step for visualizing a GIS dataset is to read the GIS data from the data file so that it can be manipulated as needed. Converting polylines into variables from a text file and then performing calculations (e.g., length) on them using the GIS data is required by the visualization process. The basic GIS data vector is geographic coordinates, which define where they will be drawn on Canvas. Many GIS data coordinates are presented in latitude and longitude measurements ($-90$ to $90$ and $-180$ to $180$), unlike computer monitor measurements ($800 \times 600$ or $1024 \times 768$

or 1600 × 1200). The monitor system is also called the monitor coordinate system. The monitor coordinate system starts from the top left as an origin (0, 0), and increases in value from top to bottom (y) and left to right (x) on the monitor.

Other data coordinate systems, such as the Universal Transverse Mercator (UTM, Grafarend 1995), will not fit on a monitor coordinate system. Moving around map features on a GIS window while panning through the map will cause the coordinate location to readjust on the computer monitor. It is to note that the real GIS coordinates are first transformed to monitor coordinates and then shown on the monitor. This allows the geographic coordinate system to assimilate to the monitor coordinate system on the computer. This section discusses how to convert from a GIS coordinate system to a computer monitor system.

Day-to-day geospatial coordinates for points, polylines, or polygons are floating numbers, and are represented in the 2D Cartesian coordinate system (Pick and Šimon 1985). The geographic coordinate system has the lower coordinate values in the bottom left and the larger coordinate values in the upper right. However, a monitor coordinate system point is represented as a pixel with integers x and y numbered from the origin in the upper left corner (Figure 5.3 right). Typically, the geographic area being analyzed is bigger than the monitor screen size, so to accurately represent the geographic data on the screen, a mapping conversion needs to be implemented. The coordinate system transformation converts point (x, y) on the geographic coordinate system (Figure 5.3 left) to the (winx, winy) on the computer monitor (Figure 5.3 right).

To conduct the coordinate transformation, calculate two critical parameters, length ratios and reference (or control) points. The length ratio is the reflecting factor from the geographic coordinate system to the monitor coordinate system. For example, using 800 in the monitor will transform to a 360 degree longitude in the geographic coordinate system. The length ratio will be 360/800. The reference point is referred to as a point in the geographic coordinate system, showing a specific point (e.g., 0, 0) in the monitor coordinate system. When initializing a map, the entire dataset of the geographic domain should be displayed. Therefore, select the upper left point of a GIS

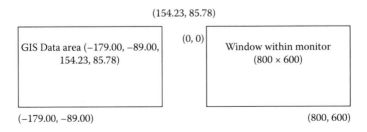

FIGURE 5.3
An example of geographic area and window monitor size with 800 × 600.

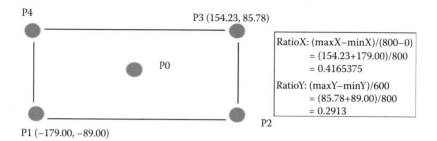

**FIGURE 5.4**
Calculation of ratioX and ratioY.

dataset (the point with the lowest x coordinate and highest y coordinate) as the origin of a computer monitor (0, 0,upper left).

In the coordinate transformation process, keep the map's direction, meaning have the monitor map with the same top to bottom and left to right direction. Along the x-axis, both the geographic coordinate system and the monitor coordinate system will increase from left to right. However, along the y-axis, the monitor coordinate system will increase from top to bottom while the geographic coordinate system will increase from bottom to top. Therefore, flip the direction when converting the coordinates from a geographic system to the monitor system. This is done by flipping the positive/negative signs of length along the y direction. The following illustrates how to calculate the coordinates corresponding to monitor coordinates (winX, winY).

Both the geographic extent and monitor screen areas have five important points (Figure 5.4) used to calculate length ratios: the center point (p0) and its four corners (p1, p2, p3, and p4). Length ratios include both ratioX, which is the length conversion ratio from geographic coordinates to monitor coordinates along the X direction, and ratioY along the Y direction. The final *ratio* used to convert geographic coordinates to monitor coordinates is selected from either ratioX or ratioY.

TIPS: Using different ratios for the x- and y-axis will cause the map to be distorted in the visualized map.

A *reference point* is another important element when converting geographic coordinates to monitor coordinates. Reference points could be centers of both systems, which are (−12.385, −1.61) and (400, 300), or upper left of both systems: (−179.00, 85.78) and (0, 0).

### 5.2.1 How to Determine Ratio Value?

Figure 5.4 shows the calculation of ratioX and ratioY based on the four corner points. Figure 5.5 shows monitor coordinates of four corner points based on ratioX and ratioY. Using p2 as an example, use the upper left corner p4 as the

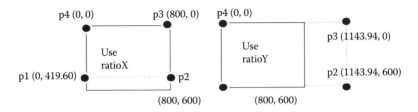

**FIGURE 5.5**
The monitor coordinates of four corner points are based on ratioX (left) and ratioY (right), using the upper left corner as the reference point.

control point and ratioX as the ratio to convert it from the geographic coordinates to monitor coordinates:

$$winX(p2) = (X(p2) - X(p4))/ \ ratioX = (154.23+179.00)/ \ 0.4165375 = 800$$
$$winY(p2) = (Y(p2) - Y(p4))/ \ ratioX = (85.78+89.00)/ \ 0.4165375 = 419.60$$

When using ratioY as the ratio, to convert it from the geographic coordinates to monitor coordinates:

$$winX(p2) = (X(p2) - X(p4))/ \ ratioX = (154.23+179.00)/ \ 0.2913 = 1143.94$$
$$winY(p2) = (Y(p2) - Y(p4))/ \ ratioX = (85.78+89.00)/ \ 0.2913 = 600$$

As shown in Figure 5.5, using ratioX will not use all 600 pixels of window height; however, not all features will show up while using ratioY (window coordinates are out of boundary). Typically, the larger one should be selected as the *ratio* value to ensure that all features are displayed on the screen at the initialization stage.

Finally, transform the geographic coordinates (x, y) to the screen pixel coordinates (winx, winy) after both the ratio and reference points (X0, Y0) are determined using the following formula:

$$winx = (X–X0)/ratioX$$
$$winy= - (X - Y0)/ratioY \ (\text{add a negative sign to flip the y-axis direction})$$

For example, if ratioX and an upper left point (–179.00, 85.78) are (0, 0), any point (x, y) from the GIS data will serve as the coordinates for the monitor window (Figure 5.6):

$$winx = (x - (–179.00))/ratioX$$
$$winy= - (y - (85.78))/ratioX$$

Therefore, the given input parameters are as follows:
   Geographic coordinate system: any point (x, y), minx, miny, maxx, maxy
   Monitor coordinate system: size (width, height) and (0, 0) at left corner

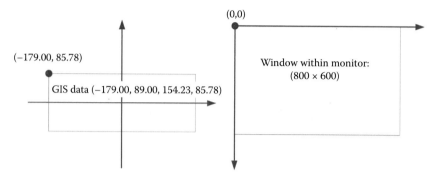

**FIGURE 5.6**
Coordinate conversion using ratioX as ratio and the upper left corner as the reference point.

The monitor coordinates of point (x, y) can be determined using the following formula:

ratioX = (maxx-minx)/width
ratioY = (maxy-miny)/height
ratio = ratioX>ratioY?ratioX:ratioY
winx = (x-(minx))/ratio
winy = -(y-maxy)/ratio

## 5.3 Visualizing Vector Data

Once the geographic coordinate system is converted into the monitor coordinate system, use the Tkinter package to visualize the data. The Canvas widget provides a general-purpose visualization background (cloth) for displaying and manipulating graphics and other drawings. The Canvas is a window itself and its origin is located in the upper left corner (0, 0). Therefore, any data visualization will display in this context starting from the upper left corner of Canvas. When creating a Canvas object, there are several important arguments, such as width (specifying the width of the Canvas window), height (specifying the height of the Canvas window), and bg (specifying the background color of the Canvas window). When drawing on Canvas, the create_xx (the second column in Table 5.2) method will construct different graphs.

Each of these methods will take different arguments as listed in the table. For example, create_arc will take the parameters of x0, y0, x1, y1, options.... The (x0, y0) point will define the upper left point and the (x1, y1) point will

**TABLE 5.2**

Canvas Widgets Can Be Created Using Different Create Methods

| Graphs | Method Context | Parameters | Usage in GIS Notes |
|---|---|---|---|
| A slice out of an ellipse. | create_arc(x0, y0, x1, y1, options) | (x0, y0, x1, y1) is the rectangle into which the eclipse, (x0, y0) and (x1, y1) are the two diagonal points | Some symbols |
| An image as a bitmap. | create_bitmap(x, y, options) | (x, y) is the point location where the bitmap is placed | Show raster images |
| A graphic image. | create_image(x, y, options) | (x, y) is the point location where the image is placed | Show raster images |
| One or more line segments. | create_line(x0, y0, x1, y1,..., options) | (x0, y0, x1, y1,...) is the list of the points in the polyline, (x0,y0) and (x1, y1) are the two diagonal points | Some polyline features such as rivers, roads |
| An ellipse; use this also for drawing circles, which are a special case of an ellipse. | create_oval(x0, y0, x1, y1, options) | (x0, y0, x1, y1) is the rectangle into which the eclipse, (x0,y0) and (x1, y1) are the two diagonal points | Some ellipse features or symbols |
| A polygon. | create_polygon(x0, y0, x1, y1,..., options) | (x0, y0, x1, y1,...) is the list of the points in the polygon | Some polygon features such as lakes, sea, cities |
| A rectangle. | create_ rectangle(x0, y0, x1, y1, options) | (x0, y0, x1, y1) is the rectangle, (x0,y0) and (x1, y1) are the two diagonal points | A map border etc. |
| Text annotation. | create_text(x, y, options) | (x, y) is the point location where the text is placed | Texture Caption |
| A rectangular window. | create_window(x, y, options) | (x, y) is the point location where the window is placed | A Canvas to draw the map |

define the lower right point of the rectangle (in which the arc will be drawn). There are many options, such as start (the beginning angle of an arc), extent (the width of the arc in degrees), and fill (the color used to fill in the arc). Figure 5.7 shows how to create a Canvas object and create three arcs using the same rectangle points, but with different colors and extents. As illustrated, the angle starts from a positive x-axis and goes counterclockwise.

As shown in Figure 5.7a, the source code creates a window, where Canvas is drawn creating a 30 degree arc (270 degrees red, 60 degrees blue, and 30 degrees green).

(a)

```
from Tkinter import*
root = Tk()
can = Canvas(root, width = 800, height = 600)
can.pack ()
xy = 20, 20, 300, 180
can.create_arc(xy, start=0, extent=270, fill="red")
can.create_arc(xy, start=270, extent=60, fill="blue")
can.create_arc(xy, start=330, extent=30, fill="green")
root.mainloop ()
```

(b)

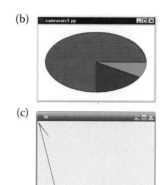

(c)

**FIGURE 5.7**
Create an arc with Tkinter. (a) Source code, (b) Draw arc, (c) Draw line.

QA: replace lines 6, 7, and 8 with the following line:

```
can.create_line (1,2,35,46,5,6,76,280,390,400)
```

Then run the code and check the GUI to see whether it is the same as Figure 5.5c.

## 5.4 Point, Polyline, Polygon

The three most popular GIS feature data types include point, polyline, and polygon. Proper methods should be selected to visualize each of them. Table 5.1 lists the methods available for visualizing different types of graphs. The polyline and polygon can be matched by create_line() and create_polyon() methods. The point can be represented by using other methods such as create_rectangle() or create_arc(). Knowing the specific size of an arc and rectangle is important when filling it in with a color, forming a circle or point. Therefore, the point, polyline, and polygon data can be visualized using create_arc(), create_line, and create_polygon, respectively.

Within the arguments of create_arc(xy, start, extent, fill=color),

- xy can be a list of [x0, y0, x1, y1, … xn, yn]
- start and extent can be the default, which is the entire circle
- fill in color can be taken as feature visualization symbols defined by users

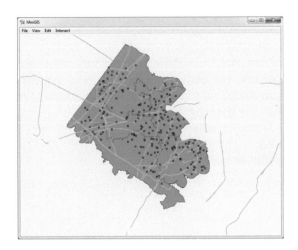

**FIGURE 5.8**
A simple Python GIS map with point, polyline, and polygon visualized.

Within the arguments of create_line(xy, options):

- xy can be a list of [x0, y0, x1, y1, x2, y2, ...]
- options can be
  - fill: to specify the color of the line
  - width: to specify the pixel width of the line

Within the arguments of create_polygon(xy, options):

- xy can be a list of [x0, y0, x1, y1, x2, y2, ...., x0, y0]. Note: the first and last points are the same
- options include
  - fill: to specify the fill color of the polygon
  - outline: specify the border line color
  - width: to specify the pixel width of the border line

Using these methods, the point, polyline, and polygon data drawn in Figure 5.1 can be visualized in Python Tkinter Canvas as Figure 5.8.

## 5.5 Programming Thinking

Writing code involves a way of thinking that is foreign to most people; however, writing code is not a unique process and is an analytical process. Practicing the following steps will improve programming skills.

### 5.5.1 Problem Analysis

The first step in programming is to analyze the problem and understand the detailed requirements. One of the problem requirements of Chapter 3 is to create a polygon and calculate the length of the polygon's border. Part of the problem requirements of Chapter 4 are to program coordinates of two polylines and calculate the length of the polylines.

### 5.5.2 Think in Programming

The second step is to think from a programming perspective, considering several factors:

- How many components should be included in the code? For example, in the Chapter 4 problem, three components are included (Figure 5.9).
- What is the sequence of these components? After components have been determined, think about the programming sequence. Which component should run first and which should follow?
- Lastly, think about how to design data and program structures, and organize the data and program structures to make them simple, reusable, and object-oriented in nature. Know how to manage object creation, compose objects into larger structures, coordinate control flow between objects, and obtain the results in the sequence obtained from the problem analyses.

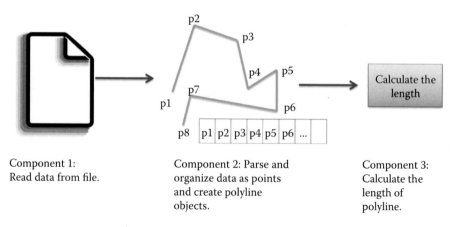

Component 1:
Read data from file.

Component 2: Parse and organize data as points and create polyline objects.

Component 3: Calculate the length of polyline.

**FIGURE 5.9**
Programming components/steps for the Chapter 4 problem.

### 5.5.3 Match Programming Language Patterns and Structure

Now, match the programming components and sequences (design of programmable components) to language structures and patterns (Proulx 2000). A program can be implemented using different programming languages and can match the components to a more familiar language structure. There are several patterns to follow to implement each component.

- *First: Functional match.* Similar to Chapter 4's file operations, reading data from the file (component 1) can open, read, and close the file in different manners. Parsing and organizing data as points and creating polyline objects is similar to parsing data from a text file (using split() and list operations), creating point objects, and creating polyline objects by defining and using Point, Polyline classes, and their composition relationship.

- *Second: Sequential match* different components. The sequence of the three components (Figure 5.5) means you should first read data, parse the data and create objects, and then calculate the polyline length. File operations are the order of opening, reading, and closing files.

- *Third: Detailed data and programming structure match.* Polyline length calculation, for example, is adding many line segment lengths through a loop. Storing a list of points uses the list data type. Parsing coordinates is using the split and data conversion for each coordinate in a loop fashion.

### 5.5.4 Implement Program

After analysis, programming with familiar language structure and patterns should work in the following order:

- Read data from the file.
- Add Point and Polyline classes and list data structures to hold the datasets before parsing the data.
- Parse data into a list of coordinates or points.
- Create point and polyline object before using polyline object.
- Call the method to calculate the polyline length.

After the first version of the program is developed,

- Again, sort through the logic of the programming, fix any remaining problems, and optimize the codes.

- Debug and test the results of the program developed.

TIPS: In the programming thinking process, there will be times where you are not as familiar with the particular pattern or structure. So, in effect, there are several ways to conduct the search:

- Search for similar patterns written in "python programming ..." by replacing the ... with the pattern or structure needed.
- Search the help document from the Python IDLE (if there is no Internet available) by typing the ... part into the search box for help.
- Discuss with peers: there are similar problematic patterns. Otherwise, post a message or question on the Python programming blog.

The more patterns practiced, the more experienced a programmer will become to find a solution.

## 5.6 Hands-On Experience with Python

### 5.6.1 Reading, Parsing, and Analyzing Text File Data

Try Code 5.3 for the problem from the last chapter and review the programming thinking process introduced in this chapter.

```
"""
```

Chapter#4

Read the following data:

Polyline;

1: 1603714.835939442,142625.48838266544; 1603749.4678153452,142620.212 43656706; 1603780.3769339535,142607.37201781105; 1603801.475846678,142 582.27024446055; 1603830.4767344964,142536.14692804776;

2: 1602514.2066492266,142330.66992144473; 1602521.4127475217,142414.92 978276964; 1602520.1146955898,142433.93817959353; 1602501.3840010355,1 42439.12358761206; 1602371.6780588734,142417.84858870413; 1602351.6610 373354,142408.02716448065; 1602334.5180692307,142388.58748627454; 160 2331.6999511716,142376.66073128115; 1602334.8067251327,142348.9653227 32; 1602338.308919772,142323.6111663878; 1602349.0226452332,142314.50 124930218; 1602363.9090971674,142310.79584660195; 1602514.2066492266, 142330.66992144473;

Code 5.3 defines the function 'readPolylineFile' to read data line by line. The readPolylineFile function will return two values: polylines and

```
>>> import math
>>> class Points:
        def __init__(self, x=0.0, y=0.0):
            self.x,self.y = x, y
>>> class Polyline:
        def __init__(self, points =[]):
            self.points = points
        def getLength(self):
            i = 0
            length =  0.0
            while i < len(self.points)-1:
                length += math.sqrt((self.points[i+1].x
                                    -self.points[i].x)**2 +
                                    (self.points[i+1].y -self.
                                    points[i].y)**2 )
                i += 1
            return length

#
## function to read out data one line by one line and
## get all points from both lines
## return two objects: points list and
## the number of the points from the first line
>>> def readPolylineFile(fileName):
        f = open(fileName, 'r')
        polylines, points, index = [], [],0
        firstPolyLineNum = 0
        for line in f:
            index += 1
            if index == 1:
                continue
            coords = line.split(':')[1]
            eachcoords = coords.split(';')
            coordsLen = len(eachcoords)
            if index == 2:
                firstPolyLineNum = coordsLen-1
                print 'The first polyline number is : ',
                firstPolyLineNum
            for i in range(coordsLen-1):
                singlecoords = eachcoords[i]
                #print 'singlecoords,', singlecoords
                xCoord = singlecoords.split(',')[0]
                yCoord = singlecoords.split(',')[1]

#print 'xCoord and yCoord, ',  xCoord, yCoord
                point = Points(float(xCoord),float(yCoord))
                points.append(point)

        f.close()
        return points, firstPolyLineNum

The first polyline number is :  5
```

**CODE 5.3**
Read from text file and create a polyline to hold data and analyze data.           (*Continued*)

```
## call the function to read data and put into points list
>>> resuts = readPolylineFile('polylinesHw4.txt')
>>> points = resuts[0]
>>> firstPolylinePointNum = resuts[1]
>>> length = len(points)
>>> print 'The total points and the numberof points for
             firstpolyline is',\
       length, firstPolylinePointNum

The total points and the numberof points for firstpolyline is 18 5
## Gets the points for first polyline and calculate length
>>> pointsForFirstPoly = points[0:firstPolylinePointNum]
>>> polyLine1 = Polyline(pointsForFirstPoly)
>>> lengthForFirstPoly = polyLine1.getLength()
>>> print "Length for first polyline -> ", lengthForFirstPoly

Length for first polyline ->  155.775923237

## Gets the points for second polyline and calculate length
>>> pointsForSecondPoly = points[firstPolylinePointNum:]
>>> polyLine2 = Polyline(pointsForSecondPoly)
>>> lengthForSecondPoly = polyLine2.getLength()
>>> print "Length for Second polyline -> ", lengthForSecondPoly

Length for Second polyline ->  549.438874589
```

**CODE 5.3** (*Continued*)
Read from text file and create a polyline to hold data and analyze data.

firstpolylineNum, which refers to how many points we have for first polyline.

### 5.6.2 Create GIS Objects and Check Intersection

a. *Problem statement*: Create four random points and four random rectangles, and find out whether any of the points are located inside any of the rectangles. Write the results to a text file.

b. *Programming thinking of the problem*: The workflow of this process is to (a) create four points, (b) create four rectangles, and (c) check each point to determine whether it is in any of the rectangles. The problem involves two data structures: Point and Rectangle classes. First, the "declare classes" patterns for Points and Rectangles will hold relevant datasets created from random module's random method.

The Point class and Rectangle class can be defined with the following attributes and functions (Figure 5.10).

*## 1. Declare the Point class*

*class Point:*

   *pass ## Implement me*

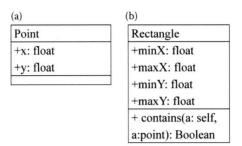

(a)

| Point |
| --- |
| +x: float |
| +y: float |
| |

(b)

| Rectangle |
| --- |
| +minX: float |
| +maxX: float |
| +minY: float |
| +maxY: float |
| + contains(a: self, a:point): Boolean |

**FIGURE 5.10**
Two classes to be created for the point and rectangle problem.

*## 2. Declare the Rectangle class*

*class Rectangle:*

> *pass ## Implement me*

The problem requires creating four random points and rectangles, which indicate that the following two steps need to be included in the program:

*## 3. Generate four points*

*points = []*

*for i in range(4): ## Loop 4 times*

> *pass ## Implement me*

*## 4. Generate four rectangles*

*rectangles = []*

*for i in range(4): ## Loop 4 times*

> *pass ## Implement me*

To check each of the four points, loop through the four points, and check to see whether each of the points is in any of the four rectangles, and then loop through the four rectangles. This will require a double loop to process.

*## 5. Check which point is in which rectangle and record the result*

*for i in range(4):*

> *for j in range(4):*

>> *#check if points[i] is in rectangles[j] and record results into a file*

>> *pass ## Implement me*

There are two components in the last check process: (a) how to check if a point is within a rectangle (the contains() method in rectangle class), and how to write the results to a file (the file open, write, and close pattern). Since the file needs to be written as you progress through the double loops,

```
>>> import random
## 1. Declare the Point class
>>> class Point:
        def __init__(self,x = 0.0, y = 0.0):
            self.x = x
            self.y = y

## 2. Declare the Rectangle class
>>> class Rectangle:
        def __int__(self):
            ## A rectangle can be determined by (minX, maxX)
            (minY, maxY)
            self.minX = self.minY = 0.0
            self.maxX = self.maxY = 1.0
        def contains(self, point): ## Check if a point is within
                                     a rectangle
            pass  ## Implement me
## 3. Generate four points
#define a Point list to keep four points
>>> points = []
#generate four points and append to the points list
>>> pass  ## Implement me

## 4. Generate four rectangles
#define a Rectangle list
>>> rects = []
>>> for i in range(4):
        rectangle = Rectangle()
        ## Generate x
        x1 = random.random()
        x2 = random.random()
        ## make sure minX != maxX
        while(x1 == x2):
            x1 = random.random()
        if x1<x2:
            rectangle.minX=x1
            rectangle.maxX=x2
        elif x1>x2:
            rectangle.minX=x2
            rectangle.maxX=x1
        ## Develop codes to generate y values
        ## and ensure the rectangle is not a line below
        pass  ## Implement me
        ## add to the list
        rects.append(rectangle)

## 5. Add code to check which point is in which rectangle
>>> resultList = []  ## And use a list to keep the results
>>> pass  ## Implement me

## 6. write the results to file
>>> f=open('HM5_Result.txt','w')
>>> for result in resultList:
        f.write(result+'\n')
>>> f.close()
```

**CODE 5.4**
Generating four points, rectangles, checking contains() relationships, and outputting results to a text file.

open it before entering the loop and close after exiting the loop. Write the file when executing the loops.

The random.random() method may generate the same values for x1 & x2 or y1 & y2 (which means a line instead of a rectangle). This can be handled by adding a method to check whether they are the same in order to prevent an invalid rectangle.

Based on this programming thinking process, the programming codes can be developed in the flow in Code 5.4:

## 5.7 Chapter Summary

This chapter introduces how to think like a programmer using GIS vector data visualization and includes

- Problem analyses
- Pattern matching
- Coordinate transformation
- Drawing vector data on Canvas
- Two coding examples are used to demonstrate the programming thinking process: (a) reading, parsing, and calculating length for polylines, and (b) generating random points and rectangles; and check the contain relationship between every point and rectangle

### PROBLEMS

1. There is a module named random in Python; import it and use its method random() to generate a random number from 0 to 1.

2. There is a popular algorithm in GIS to find whether a point is inside a rectangle based on their respective point coordinates (x, y, and minx, miny, maxx, maxy). Describe the algorithm in a mathematical algorithm using (x, y, and minx, miny, maxx, maxy).

3. Write a program to (a) generate m number of points and n number of rectangles (m and n can be changed through user input), (b) check which points are in which rectangles.

4. Program to write the point coordinates and rectangles point coordinates to a text file, and then write the result of (2) to the text file.

5. In a Word document, explain the "point in rectangle" algorithm and program created, and code the program in a .py file to find which point generated in (3) is within which rectangle generated in (3). Then check the text file output.

# 6

## Shapefile Handling

One of the most important functions of GIS software is to read popular GIS data file formats, such as shapefiles. A shapefile is a binary data file format originally developed by ESRI, and has been widely used for exchanging vector data among different GIS professionals and communities (ESRI 1998). This chapter introduces how shapefiles are formatted and how to read them with Python, that is, reading binary data, reading a shapefile header, reading a point shapefile, and reading polyline and polygon shapefiles.

### 6.1 Binary Data Manipulation

Shapefiles are in binary format and the Python module for manipulating binary data is known as *struct*. The *struct* module has a number of functions to read data from and write data into binary format. This section introduces how we can use the *struct* module to handle binary data.

*Struct* handles binary data by converting data back and forth between its original format and binary format. The pack function of *struct* is used to convert data into a binary format. For example, the following statement returns a binary string containing the values v1, v2, ... packed according to the given format *fmt*. The arguments must match the values required by the format exactly.

   *struct.pack(fmt, v1, v2, ...)*

The unpack function of *struct* is used to interpret binary data to its original value (Code 6.1). For example, the following statement unpacks the binary (presumably packed by *struct.pack(fmt, v1, v1, ...)*) according to the given format *fmt*. The result is a tuple, even if it contains only one element. The binary data must contain exactly the same number of data as required by the format, that is, len(binary data) must equal calcsize(fmt).

   *struct.unpack(fmt, binarydata)*

The *struct* module must be imported before using (the first statement of Code 6.1). The code also demonstrates how to pack two integers (100, 200) represented by variables (x, y) into a binary string. String 'ii' is used to represent two integer values with each 'i' representing one integer. The fifth

```
import struct
x,y = 100,200
s = struct.pack('ii',x,y)
print s
result = struct.unpack('ii',s)
print result
```

**CODE 6.1**
Examples of pack and unpack methods of the struct module.

**TABLE 6.1**

Format Characters

| Format Character | C Type | Python Type | Standard Size |
|---|---|---|---|
| C | char | string of length 1 | 1 |
| B | signed char | Integer | 1 |
| B | unsigned char | Integer | 1 |
| ? | _Bool | Bool | 1 |
| h | short | Integer | 2 |
| H | unsigned short | Integer | 2 |
| i | int | Integer | 4 |
| I | unsigned int | Integer | 4 |
| l | long | Integer | 4 |
| L | unsigned long | Integer | 4 |
| q | long long | Integer | 8 |
| Q | unsigned long long | Integer | 8 |
| f | float | Float | 4 |
| d | double | Float | 8 |

```
>>> import struct
>>> i,b,x,y,z = 100,True,-180.0,90,0.212
>>> s = struct.pack('<ibdfd',i,b,x,y,z)
>>> len(s)
>>> result = struct.unpack('<i?dfd',s)
>>> print result
(100, True, -180.0, 90.0, 0.212)
```

**CODE 6.2**
Packing and unpacking different data types using proper format.

statement unpacks the binary string into its original data value (100, 200). The string 'ii' is important and referred to as the string *format* (denoted as *fmt*), which is used to specify the expected format, and is required to call both pack and unpack methods. Table 6.1 details the format characters used in the string *fmt* to specify format for packing and unpacking binary data.

Code 6.2 shows how to pack different formats using the format characters. Five variables representing five values in data types of integer, Boolean,

**TABLE 6.2**

Struct Starting Character

| Character | Byte Order | Size | Alignment |
|-----------|-----------|------|-----------|
| @ | native | native | native |
| = | native | standard | none |
| < | little-endian | standard | none |
| > | big-endian | standard | none |
| ! | network (= big-endian) | standard | none |

double, float, and double are packed. The total length of the packed string is 4(i) + 1(b) + 8(d) + 4(f) + 8(d) = 25. Because the *struct* package is following C standard, the C Type is used. Python has fewer data types; however, the standard size of each data type can be kept if the first character of the format string is indicated by the byte order, size, and alignment of the packed data.

By default, the @ will be used if the first character is not one of the characters given in Table 6.2 below:

- Native byte order is big-endian or little-endian, depending on the host system. For example, Intel x86 and AMD64 (x86-64) are little-endian; Motorola 68000 and PowerPC G5 are big-endian; ARM and Intel Itanium feature switchable endian-ness (bi-endian). Use sys. byteorder to check the endian-ness of the system.

- Native size and alignment are determined using the C compiler's *sizeof* expression. This is always combined with native byte order.

- Standard size depends only on the format character as defined in Table 6.1 above.

- Note the difference between '@' and '=': both use native byte order, but the size and alignment of the latter is standardized.

- The form '!' is used when we cannot remember whether network byte order is big-endian or little-endian.

Byte order[*] concept is also used in the ESRI Shapefile format. Big-endian and little-endian byte orders are two ways to organize multibyte words in the computer memory or storage disk. When using big-endian order, the first byte is the biggest part of the data, whereas the first byte is the smallest part of the data when using little-endian order. For example, when storing a hexadecimal representation of a four-byte integer $0 \times 44532011$

---

[*] "Endianness." Wikipedia. Wikipedia Foundation, Inc., http://en.wikipedia.org/wiki/Endianness.

**TABLE 6.3**

Four-Byte Integer 0 × 44532011 in the Storage

| | |
|---|---|
| Big-Endian | 44 53 20 11 |
| Little-Endian | 11 20 53 44 |

```
import struct
x,y,z = 100,200,300
s = struct.pack('>iii',x,y,z)
print s
result = struct.unpack('>iii',s)
print result
result = struct.unpack('<iii',s)
print result
```

**CODE 6.3**
Pack and unpack must use the same byte order.

```
#This code should be typed in interactive window
import struct
x,y,z = 100, 200, 300
s = struct.pack('<iii',x,y,z) #little_endian
coding
s
'd\x00\x00\x00\xc8\x00\x00\x00,\x01\x00\x00'
s = struct.pack('iii',x,y,z) #default coding
s
'd\x00\x00\x00\xc8\x00\x00\x00,\x01\x00\x00'
```

**CODE 6.4**
The default byte order is little-endian for our computers.

(Table 6.3) using the big-endian order binary format, the byte sequence would be "44 53 20 11."

- For big-endian byte order, use '>' while packing or unpacking the bytes, for example, struct.unpack('>iiii', s).
- For little-endian byte order, use '<', for example, struct.unpack('<iiii', s).
- If these two are mixed, for example, packing using '>' and unpacking using '<', an unexpected result will be generated, as shown in the last statement of Code 6.3.

While constructing the formatted string, the byte order should be specified. In the absence of a byte order symbol, *packing and unpacking* will use little-endian by default for PCs (Code 6.4). The results of two codes show no difference. The results are hexadecimal, that is, 0x 00000064 represents 100, 0x 0000c8 represents 200, and x0000012c represents 300. More about *struct* and formatting info can be found at http://docs.python.org/library/struct.html.

## 6.2 Shapefile Introduction

A shapefile includes multiple files with the same file name and different file extensions. Each vector dataset is stored in several files, such as .shp (main file), .shx (index file), .dbf (dBASE file), .prj (projection file), etc. (Figure 6.1). The main file, index file, and dBASE file are required by all GIS software that works with shapefiles. The triple must have the same name, for example, mymap.shp, mymap.shx, and mymap.dbf:

- .shp: The .shp file contains the vertices of all entities (Figure 6.1). The vertices are organized hierarchically in features/records, parts, and points. The .shp file also contains information on how to read the vertices (i.e., as points, lines, or polygons). Some important attributes can also be termed as the third dimension (measurements), and stored in the .shp file.
- .shx: An index is kept for each record, and is beneficial for finding the records more quickly.
- .dbf: Attribute information is stored in the .dbf file associated with each .shp file. The .dbf file contains dBASE tables and stores additional attributes that cannot be kept in a shapefile's features. It contains exactly the same number of records as there are features in the .shp file (otherwise the data could not be interpreted). The records belong to the shapes sequentially, meaning that the first, second, and third records belong, respectively, to the first, second, and third, features in the .shp file. If we edit the .dbf using a third-party tool and alter the records, the order may be lost. More information can be found from the ESRI shapefile format white paper (ESRI 1998).

There are 14+ types of features supported by shapefiles, such as point, polyline, and polygon. 3D features are added to shapefile structure with a dimension dedicated to z value.

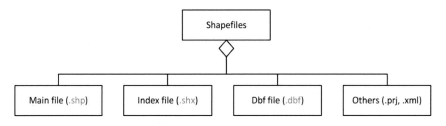

**FIGURE 6.1**
Shapefile structure. (Adapted from ESRI. 1998. ESRI Shapefile Technical Description. An ESRI White Paper, 34.)

## 6.3 Shapefile Structure and Interpretation

### 6.3.1 Main File Structure of a Shapefile

The main file contains a file header and all feature records (Figure 6.2). Each feature record is composed of a record header and record contents.

#### 6.3.1.1 Main File Header

Figure 6.3 shows the file header format including 100 bytes. The first 4 bytes indicate this file is a shapefile with an integer value of 9994. Bytes 4 to 24 are integer 0 and reserved for future use. The first 28 bytes, including the data description fields of the file header and file length (integer in bytes 24–27), are

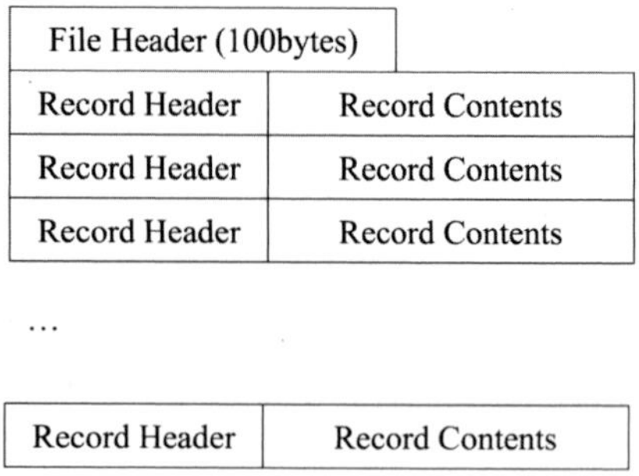

| File Header (100bytes) | |
|---|---|
| Record Header | Record Contents |
| Record Header | Record Contents |
| Record Header | Record Contents |

. . .

| Record Header | Record Contents |
|---|---|

**FIGURE 6.2**
Shapefile main file structure.

**Shape Types**

| Position | Field | Value | Type | Byte Order |
|---|---|---|---|---|
| Byte 0 | File Code | 9994 | Integer | Big |
| Byte 4 | Unused | 0 | Integer | Big |
| Byte 8 | Unused | 0 | Integer | Big |
| Byte 12 | Unused | 0 | Integer | Big |
| Byte 16 | Unused | 0 | Integer | Big |
| Byte 20 | Unused | 0 | Integer | Big |
| Byte 24 | File Length | File Length | Integer | Big |
| Byte 28 | Version | 1000 | Integer | Little |
| Byte 32 | Shape Type | Shape Type | Integer | Little |
| Byte 36 | Bounding Box | Xmin | Double | Little |
| Byte 44 | Bounding Box | Ymin | Double | Little |
| Byte 52 | Bounding Box | Xmax | Double | Little |
| Byte 60 | Bounding Box | Ymax | Double | Little |
| Byte 68* | Bounding Box | Zmin | Double | Little |
| Byte 76* | Bounding Box | Zmax | Double | Little |
| Byte 84* | Bounding Box | Mmin | Double | Little |
| Byte 92* | Bounding Box | Mmax | Double | Little |

* Unused, with value 0.0, if not Measured or Z type

0 Null Shape
1 Point
3 PolyLine
5 Polygon
8 MultiPoint
11 PointZ
13 PolyLineZ
15 PolygonZ
18 MultiPointZ
21 PointM
23 PolyLineM
25 PolygonM
28 MultiPointM
31 MultiPatch

**FIGURE 6.3**
File header of shape main file.

in big-endian byte order, and '>' is required to unpack the bytes, for example, *struct.unpack('>iiiiiii', s)*. The unit for the total length of the shapefile is 16 bit word, that is, the total file length in bytes would be double the value of the interpreted number.

The rest of the file header is in little-endian byte order, and '<' is required to unpack them, for example, *struct.unpack('<iiii', s)*. Omit the '<' since it is the default value for pack or unpack on most PCs. Starting byte 28, a 4-byte integer (value of 1000) refers to the version of the shapefile. Starting byte 32, a 4-byte integer, indicates the feature shape type (e.g., 1 means the file is for Point feature, and 3 indicates it is a Polyline file, Figure 6.3 right). Byte 36 to the 100 is the bounding box of the entire dataset in the shapefile. The bounding box includes four dimensions x, y, z, and m. Each dimension includes minimum and maximum values in the sequence of minx, miny, maxx, maxy, minz, maxz, minm, and maxm. The bounding box should be written with fmt '<ddddddddd' for all values in double data type.

*Hands-on practice*: Interpret the shapefile header (Code 6.5)

- Open the Python GUI console.
- Copy the data given including Schools.shp, Schools.shx, Schools.dbf to c:\code\data\.
- Practice the Code 6.5 with the Python interpreter GUI.

The first three statements have 28 bytes from the shapefile. The fourth statement unpacks the data in big-endian order and has seven integers. The first integer is 9994, the file code for shapefiles. The next five are zero, and are reserved for future use. The last one is the file length, which is the total length of the shape main file in 16-bit or two bytes unit. Therefore, the actual total length of the file is double the indicated value (i.e., 288 * 2 = 576) bytes. The next statement reads out 72 bytes and unpacks them using little-endian byte order to obtain two integers and eight double values. The first one, with a value of 1000, is the version of shapefile. The second one, with value 1, indicates that the shapefile feature type is Point. The following four refer to

```
>>> import struct
>>> f = open('Schools.shp','rb')
>>> s = f.read(28)
>>> b = struct.unpack('>iiiiiii',s)
>>> print b
(9994, 0, 0, 0, 0, 0, 288)
>>> s = f.read(72)
>>> b = struct.unpack('<iidddddddd',s)
>>> print b
(1000, 1, 1847318.8628035933, 765532.64196603,
1859639.8841250539, 778092.9935274571, 0.0, 0.0, 0.0, 0.0)
```

**CODE 6.5**
Interpreting the shapefile main file header.

minx, miny, maxx, maxy, respectively, and minz, maxz, minm, maxm are all zeros since these two dimensions are not used in this shapefile.

### 6.3.1.2 Feature Record

Figure 6.4 shows the record header and content structure for a Point shapefile. The record header is the same for all shape types while record content is different for different feature types and specific features. The record header includes 8 bytes, where the first 4 bytes indicate the sequential record number and the next 4 bytes indicate content length for the record. The record header is in big-endian byte order. The record content for each point feature includes 20 bytes, with the first 4 bytes representing feature type, which is 1 for all points, 8 bytes for the x coordinate, and 8 bytes for the y coordinate (Figure 6.5). The record content is in little-endian byte order for Point record content.

*Hands-on practice*: Interpret the point shapefile (Code 6.6)

- Open the Python file editing window.
- Copy the data, including Schools.shp, Schools.shx, Schools.dbf into c:\code\data\.
- Type the following code in the Python editor window.

## Record Header

| Position | Field | Value | Type | Byte Order |
|----------|-------|-------|------|------------|
| Byte 0 | Record Number | Record Number | Integer | Big |
| Byte 4 | Content Length | Content Length | Integer | Big |

## Record Content for Point

| Position | Field | Value | Type | Number | Byte Order |
|----------|-------|-------|------|--------|------------|
| Byte 0 | Shape Type | 1 | Integer | 1 | Little |
| Byte 4 | X | X | Double | 1 | Little |
| Byte 12 | Y | Y | Double | 1 | Little |

**FIGURE 6.4**
Point record header and content.

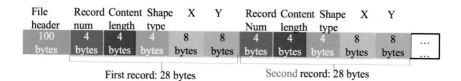

First record: 28 bytes          Second record: 28 bytes

**FIGURE 6.5**
File structure for Point .shp file.

```
>>> import struct
>>> f = open('Schools.shp','rb')
>>> f.seek(24)
>>> s = f.read(4) #Get the file length
>>> b = struct.unpack('>i',s) ##The file length is big-endian order
                                integer
>>> featNum = (b[0]*2-100)/28 ##Calculation the feature numbers
>>> out = open('schools_shp.txt','w')
>>> for i in range(featNum):
        f.seek(100+i*28+12)
        s = f.read(16)  ##16 bytes with 8 bytes for both x and y
        x,y = struct.unpack('dd',s) ##little-endian order by default
        out.write(str(i)+':'+str(x)+','+str(y)+'\n')
>>> f.close()
>>> out.close()
```

**CODE 6.6**
Read the point shapefile and write the results to a text file.

Code 6.6 reads the shapefile to get the file length (bytes 24–27 in file header) and uses the file length to calculate the number of points in this shapefile in the following steps:

1. Doubling the size to convert from 16-bit (two bytes) unit to 8-bit (one byte) unit
2. Subtracting 100 bytes for a file header
3. Dividing by 28 (each record header and record content takes 28 bytes in point shapefile) to get the feature number

The file length and number of point features are then printed out and a text file is opened to write the results. A *for* loop is used to cycle through each record/feature to read out the x, y values and print out and write to the text file. Lastly, the two files are closed to conclude the file read/write process. In the *for* loop, the first line moves the file pointer to the position where the ith record's x value starts (100 + 12 + i*28, 12 refer to the record header [8 bytes] and the shape type integer 1 [4 bytes]), then reads 16 bytes for x, y and unpacks them into x, y variables.

### 6.3.2 Index File Structure (.shx)

Figure 6.6 shows the file structure for the .shx file. The index file header is the same as the main file header, including 100 bytes and can be interpreted using Code 6.7. The file length stored in the index file header is the total length of the index file in 16-bit words (the 50 16-bit words of the header plus 4 times the number of records). This can also be used to calculate the number of records. The ith record in the index file stores the offset and content length for the ith record in the main file. The offset of a record in the main file is the number of 16-bit words from the start of the main file to the first byte of the

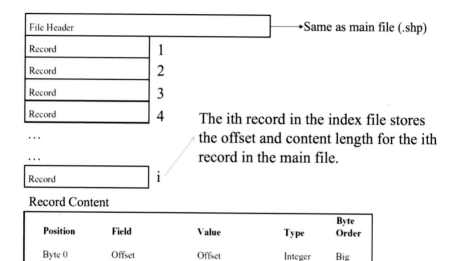

The ith record in the index file stores the offset and content length for the ith record in the main file.

**Record Content**

| Position | Field | Value | Type | Byte Order |
|----------|-------|-------|------|------------|
| Byte 0 | Offset | Offset | Integer | Big |
| Byte 4 | Content Length | Content Length | Integer | Big |

**FIGURE 6.6**
File structure for .shx file.

```
>>> import struct
#Open in binary mode for portability
>>> f = open('Schools.shx','rb')
>>> f.seek(24)
>>> s = f.read(4)
>>> b = struct.unpack('>i',s)
>>> featNum = (b[0]*2-100)/8
>>> out = open('schools_index.txt','w')
>>> for i in range(featNum):
        f.seek(100+i*8)
        s = f.read(8)
        off,length = struct.unpack('>ii',s)
        out.write(str(i)+':'+str(off)+','+str(length)+'\n')
>>> f.close()
>>> out.close()
```

**CODE 6.7**
Interpreting the shape index file to get the number of records, and the offset, content length of each record.

record header for the record. Thus, the offset for the first record in the main file is 50, given the 100-byte header.

*Hands-on practice*: Interpret the point .shx file

- Open the Python file editor window.
- If the folder "c:\code\data\" does not have the following files Schools. shp, Schools.shx, Schools.dbf, copy the data into the folder.
- Type the Code 6.7 on the Python editor.

Code 6.7 first reads the integer of index file length from bytes 24–27. This number, in 16-bit or two-byte unit is then used to calculate the feature number by

1. Doubling the value to obtain the length in byte unit
2. Subtracting 100 index file header
3. Dividing by the number of bytes for each record in index file (8 bytes)

The feature number and file length are printed and a text file is opened to keep each feature offset and content length value in the main file. The *for* loop reads each record and writes it in the text file. Again, both files are closed at the end of the program. The *for* loop cycles through each record to (a) move to the start of the ith record (100 + i*8), (b) read out 8 bytes for two integers, and (c) unpack the two integers as offset and contentLength, which are printed and written to the text file following the sequence of the record. The text file and the output on the interactive window can be used to verify the content.

### 6.3.3 The .dbf File

The .dbf file keeps attributes for the related shapefile. It has the same name as the main and index file. Each feature must have a record in .dbf file. Each feature must be in the same order in the .dbf file as it appears in the main and index files. Details of the .dbf file format can be found online* and will not be discussed in this book.

## 6.4 General Programming Sequence for Handling Shapefiles

Reading the shapefile is the first step to process the features (e.g., visualize the features on the GUI, make a map, and perform spatial analysis). There are several steps to reading a shapefile:

1. Open file to read in binary mode.
2. Read index file header and interpret the meta-information, for example, bounding box, and number of records.
3. Read records' meta-information, such as offset, and content length for each record.

---

* "Data File Header Structure for the dBASE Version 7 Table File." dBase. http://www.dbase. com/KnowledgeBase/int/db7_file_fmt.htm.

4. Read data dynamically based on each record content structure for specific shape types.
5. Assemble data into objects of point, polyline, and polygon or other types.
6. Analyze and process data as needed.
7. Prepare data to be written to a file.
8. Write formatted data to files.
9. Close files.

Steps 7 and 8 are for shapefile output. Unless converting each feature's spatial attributes (coordinates) or nonspatial attributes from a different format (e.g., a text file) to generate new shapefiles, these two steps are not required. This conversion could also be easily accomplished using ArcGIS Python scripting, which will be introduced later in this book.

## 6.5 Hands-On Experience with Mini-GIS

### 6.5.1 Visualize Polylines and Polygons

- Download and unzip the Mini-GIS package.
- Double click on the MiniGIS.py file.
- In the MiniGIS window, select File->load .shp->select the countries folder (Figure 6.7a).

(a)                                          (b)

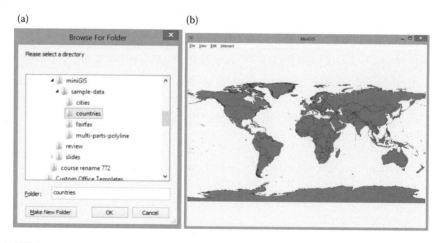

**FIGURE 6.7**
(a) Select the 'countries' folder to add data, and (b) the map window displays countries as polygons and with country borders as polylines.

The country polygons and country border polylines will be displayed (Figure 6.7b). We can operate the map by selecting View-> [Zoom In |Zoom Out|Zoom Extent|Zoom Full]. All the source codes for the file read, map display, and map operation are included in the package. Please try to identify from the Mini-GIS package the files that read shapefiles and visualize polylines and polygons.

## 6.5.2 Interpret Polyline Shapefiles

The polygon shapefiles are in the same format as polyline except that the shape type is 5 and the first point of a polygon part is the last point of the same polygon part (Figure 6.8). Therefore, the practice in this subsection can be utilized for reading polygon shapefiles with minor updates.

Reading a polyline file can leverage the code developed for reading the feature number, record offset, and record content length (Code 6.8). Once the offset and content length for each record is identified, the rest will go to the starting position of a specific record to begin interpreting the polyline. For each record header, the first integer is the record sequence number and the second integer is the content length, as shown in Figure 6.8. In the record content, the first value is an integer showing the shape type as a value 3-polyline. The following 32 bytes are the bounding box for minx, miny, maxx, maxy for the current feature. Because a polyline may include multiple lines, each line of a polyline is considered a part. The number of parts of a polyline follows the bounding box as an integer, followed by the number of points as an integer. Based on the part number, there are 4*part-Number bytes holding the starting point of each part in the entire point sequence in the current feature. The starting number is 0 for the first part. Following the part index are the points' x, y coordinates. If z and m dimensions are supplied in the format, relevant fields, such as bounding box and points coordinate values will adjust according to store z and m values. Based on these analyses, the Python code for reading a shapefile can be structured as follows (Code 6.8):

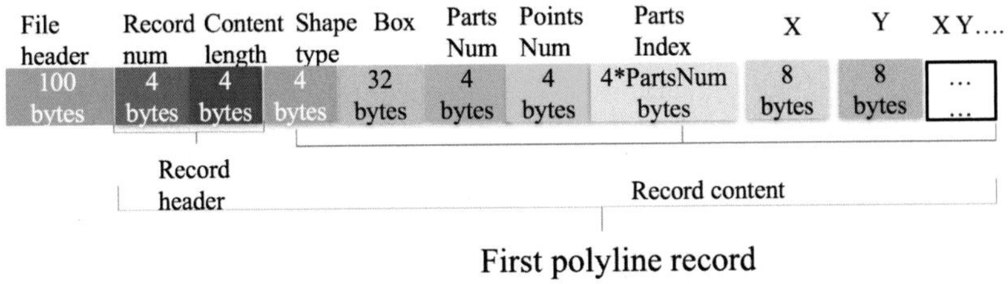

First polyline record

**FIGURE 6.8**
The shapefile polyline format with the file header and first polyline shown. The rest of the shapefile repeats the polyline record format.

```python
# import modules for the line length calculation,
# binary data unpack, and visualization.
import math
from Tkinter import *
import struct

# define point, polyline classes
class Point:
    def __init__(self, x = 0.0, y = 0.0):
        self.x = x
        self.y = y
class Polyline:
    # define object initialization method
    ## partsNum
    def __init__(self, points= [], partsNum = 0):
        self.points = points
        self.partsNum = partsNum

#-----Part 1: read and process the first 100 bytes
# #1. open index file to read in binary mode
shxFile = open("Partial_Streets.shx","rb")
# shapefile name can be replaced with any polyline

##2. read index file header and interpret the meta information, e.g.,
bounding box, and # of #records
# read first 28 bytes
s = shxFile.read(28)
# convert into 7 integers
header = struct.unpack(">iiiiiii",s)
# get file length
fileLength = header[len(header)-1]
# calculate polyline numbers in the shape file based on index file
length
polylineNum = (fileLength*2-100)/8
print 'fileLength, polylineNum:',fileLength, polylineNum
# read other 72 bytes in header
s = shxFile.read(72)
# convert into values
header = struct.unpack("<iidddddddd",s)
# get boundingbox for the shape file
minX, minY, maxX, maxY = header[2],header[3],header[4],header[5]
##3. read records¡¯ meta information, such as offset,
##    and content length for each record,

# define an empty list for holding offset of each feature in main file
recordsOffset = []
# loop through each feature
for i in range(0,polylineNum):
    # jump to beginning of each record
    shxFile.seek(100+i*8)
```

**CODE 6.8**
Reading and visualize polyline shapefiles.                    *(Continued)*

```
    # read out 4 bytes as offset
    s = shxFile.read(4)
    offset = struct.unpack('>i',s)
    # keep the offset in the list
    print 'offset is:', offset
    recordsOffset.append(offset[0]*2)
# close the index file
print recordsOffset

#--------Part 2: read each polyline and prepare them in right order.
# open the main file for read in binary
shpFile = open("Partial_Streets.shp","rb")
# shapefile name can be replaced with any polyline
# define an empty list for polylines
polylines = []
# loop through each offset of all polylines
##4. read data dynamically based on each record content structure
##   for specific shape types
for offset in recordsOffset:
    # define two lists for holding values
    x, y = [], []
    # jump to partsNum and pointsNum of the polyline and read them out
    shpFile.seek(offset+8+36)
    s = shpFile.read(8)
    # generate an empty polyline object
    polyline = Polyline()
    partsNum, pointsNum = struct.unpack('ii',s)
    polyline.partsNum = partsNum
    print 'partsNum, pointsNum: ',partsNum, pointsNum

# read the list of parts holding the starting sequential number of
point
    # in that part
    s = shpFile.read(4*partsNum)
    """
    Compose the unpack format based on number of parts
    When we unpack a binary string, we need a format (e.g., 'i' for
    one integer,
    'ii' for two integer). However, we do not know how many
    integer(partsNum)
    we need to unpack, therefore we use a loop to iterate the
    partsNum.
    For each partsNum, we add one 'i' to the str. Therefore if the
    partsNum
    equal to, for example, 2, the str will equal to 'ii' after the
    loop
    """
    str = ''
    for i in range(partsNum):
        str = str+'i'
    print 'str is :', str
```

**CODE 6.8** (*Continued*)
Reading and visualize polyline shapefiles.                    (*Continued*)

```
# get the starting point number of each part and keep in a
partsIndex list
polyline.partsIndex = struct.unpack(str,s)
# loop through each point in the polyline
points = []
for i in range(pointsNum):
    # read out polyline coordinates
    # and add to the points' x, y coordinates' lists
    # ADD CODES TO READ THE COORDINATES VALUES HERE
    #5. assemble data into objects of point, polyline,
    #    and polygon or other types.
    point = Point(x, y)
    points.append(point)
# assign points lists to the polyline
polyline.points = points
# add the polyline read to the
polylines.append(polyline)
#-------------Part 3: prepare to visualize the data
# create main window object
#8. Analyze and process (visualize) data as needed
root = Tk()
# define window size
windowWidth, windowHeight  = 800, 600
# calculate ratios of visualization
ratiox = windowWidth/(maxX-minX)
ratioy = windowHeight/(maxY-minY)
# take the smaller ratio of window size to geographic distance
ratio = ratiox
if ratio>ratioy:
    ratio = ratioy
# create canvas object
can = Canvas(root, width = 800, height = 600)
# loop through each polyline
for polyline in polylines:
    #define an empty xylist for holding converted coordinates
    xylist = []
    # loop through each point
    # and calculate the window coordinates, put in xylist
    for point in polyline.points:
        pass
# ADD CODES HERE TO TRANSFORM THE COORDINATE SYSTEM BASED ON RATIO #
FOUND
for k in range(polyline.partsNum):
        #get the end sequence number of points in the part
        if (k==polyline.partsNum-1):
            endPointIndex = len(polyline.points)
        else:
            endPointIndex = polyline.partsIndex[k+1]

        #define a temporary list for holding the part coordinates
        tempXYlist = []
```

**CODE 6.8** (*Continued*)
Reading and visualize polyline shapefiles.                                    (*Continued*)

```
        #take out points' coordinates for the part
        #and add to the temporary list
        for m in range(polyline.partsIndex[k], endPointIndex):
            pass
#ADD CODES HERE TO COMPOSE THE XYlist FOR DRAWING EACH LINE SEGMENT.
        # create the line
        #can.create_line(tempXYlist,fill='blue')
#add lines to window and show up the window
can.pack()
root.mainloop()

#9. close file
shxFile.close()
shpFile.close()
```

**CODE 6.8** (*Continued*)
Reading and visualize polyline shapefiles.

1. Type/copy the code into the programming window, save the python file.
2. Copy the data Partial_Streets.shp and Partial_Streets.shx to the same folder where you save the python .py file.
3. Run the python file.
4. Explore and analyze the code to understand each section of the code.

## 6.6 Chapter Summary

This chapter introduces vector data file formats, ESRI's shapefile format, and explains how to read the files through the Mini-GIS:

- Learn how to process binary data.
- Become familiar with different shapefile file structures, including .shp and .shx.
- Read the .shp and .shx using Python.
- Handle point shapefiles.
- Handle polyline shapefiles.

## PROBLEMS

1. Pick a polyline shapefile with 10–100 features and one polygon file with 10–100 features.

2. Write a Python program to read the polylines and polygons from the shapefiles. (We can use .shp and .shx files.)

3. Construct polyline and polygon objects in Python.

4. Draw the polylines/polygons using Tkinter package.

5. Comment and explain the program to show the logic flow of your programming by referring to Code 6.8.

# 7

## Python Programming Environment

There are several branches of Python programming environment, most of which are based on the Python Software Foundation open-source Integrated Development Environment (IDE). This chapter introduces the general aspects of the open-source IDE, including different aspects of programming interfaces, path setup, debugging, highlighting, and module management. This was exemplified in the Mini-GIS package in the previous chapters.

### 7.1 General Python IDE

Python is an interpretive (or interactive) language. Therefore, we do not need to compile the code to an executable file (e.g., .exe, or .jar file) to run it. There are two ways for a Python interpreter to read and apply the code.

#### 7.1.1 Python Programming Windows

##### 7.1.1.1 Command-Line GUI

In the windows system, the command-line GUI of Python IDE can be accessed from Windows Program→Python→Python GUI (command-line). The command-line can be used as a DOS or Linux command-line GUI. A Python statement typed in the command-line GUI will be executed directly. There are several Python commands that can only be executed on the command-line GUI (Figure 7.1), such as trying the fibo method in Chapter 6 of Python's online tutorial. However, this is the least friendly development window because you cannot customize it and the code is not colored to illustrate the code structure while typing.

##### 7.1.1.2 Interactive GUI

Another Python IDE is the Python IDLE window, where the Python interpreter will execute each command typed in the interactive GUI (Figure 7.2). IDLE is useful for practicing the Python syntax, particularly for novice users who are not familiar with the Python language, or for those who want to use Python for simple 'calculator' operations, for example, calculate $5^{10}$. When using the IDLE interface, however, the commands are lost once the window

**FIGURE 7.1**
Python Command-line GUI.

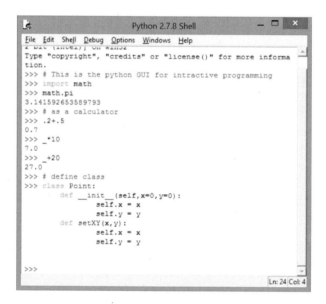

**FIGURE 7.2**
Python Interactive GUI.

is closed; therefore, the interactive GUI is not appropriate for writing complex programs.

### 7.1.1.3 File-Based Programming

A file-based programming window (can be brought up from Python IDLE through menu File→New) provides a convenient way to preserve the code by saving the code in a Python file (Figure 7.3). This window can be invoked

```
Python 2.7.8: Map.py - C:\Users\Phil\Documents\courses\14ipgis\package\Map.py    –  □  ×

File  Edit  Format  Run  Options  Windows  Help

from Layer import *
import Tkinter

class Map:
    def __init__(self, winWidth, winHeight):
        self.root=Tkinter.Tk()
        self.root.title("My Python Mini-GIS")
        self.root.columnconfigure(0, weight=1)
        self.root.rowconfigure(0, weight=1)
        self.windowWidth, self.windowHeight = winWidth, winHeight
        self.can = Tkinter.Canvas(self.root, height = self.windowHeight, width =
        #self.can.grid(column=0, row=0, sticky=(N, W, E, S))
        self.addEvents()
        self.zoomFactor = 1
        self.addButtons()
        self.controlPoint = 2 #TOPLEFT:1, CENTER:2, LOWERLEFT:3, TOPRIGHT:4,LOWE
        self.layers = []
        self.bbxset = False # bounding box
        self.ratio=1
        self.mapMode = 0 #no event: 0; drawPoint: 1; pan: 2; drawLine: 3; drawPo

    def calculate(self): #determines ratio, pick bigger one; do it each time a n
        ratiox = self.windowWidth/(self.maxx-self.minx)
        ratioy = self.windowHeight/(self.maxy-self.miny)
        self.ratio = ratiox
        if self.ratio>ratioy:
            self.ratio = ratioy
        self.ratio/=self.zoomFactor

    def addLayer(self, fileName,color):#creates layer with file name, appends to
        layer = Layer(fileName, color)
        if layer != 0:
            self.layers.append(layer)
            if self.bbxset: #bounding box for map is reset each time a layer is
                if self.minx > layer.minx:
                    self.minx = layer.minx
                if self.miny > layer.miny:
                    self.miny = layer.miny
                if self.maxx < layer.maxx:
                    self.maxx = layer.maxx

                                                                    Ln: 1 Col: 0
```

**FIGURE 7.3**
Python file-based programming window.

from the 'File→New Window' of the Python IDLE. Within this window, there are three ways to execute the code: (a) press F5, (b) click on Run→Run Module, or outside the window (c) double click on the .py file in Windows explorer (the Python IDE must be installed for this to work).

### 7.1.2 Python IDE Settings

The Python IDLE can be customized with the "IDLE Preferences" dialog, which can be invoked through Options→Configure IDLE in the IDLE menu.

#### 7.1.2.1 Highlighting

Coloring the code can help you better understand, capture, communicate, and interact with peer programmers. In Python IDLE, code (including comments) can be highlighted with different colors based on the types of the

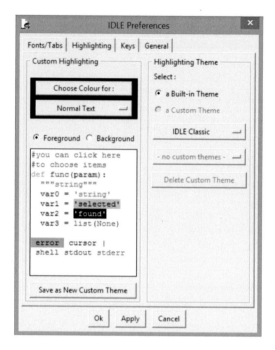

**FIGURE 7.4**
Color of different parts of a program can be highlighted for better formatting, communication, interaction, and programming.

words input. For example, keywords can be set as yellow, variables as black, and functions as blue (Figure 7.4). These settings can be customized in the "Highlighting" tab.

### 7.1.2.2 General Setting of the Programming Window

The initial status of Python IDLE is configured in the "General" tab (Figure 7.5). Settings include which window to initialize (either the interactive or the edit window), how to save the code, the initial window size, paragraph formatting, and source encoding method.

### 7.1.2.3 Fonts Setup for the Coding

The "Fonts/Tabs" tab can set up the code font size, style, and indentation width (Figure 7.6).

### 7.1.3 Debugging

Testing the code to fix errors and improve the robustness is called debugging. It is a must-have process in programming because it is almost impossible to

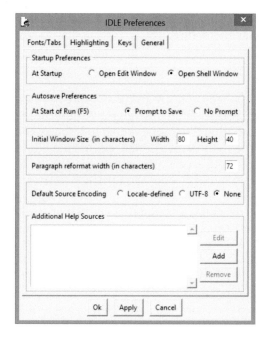

**FIGURE 7.5**
General setting for the initial status of Python IDLE.

**FIGURE 7.6**
Customize font size, style, and indentation for Python IDLE.

write bug-free codes. This section will go through basic skills for errors/exceptions handling and debugging.

### 7.1.3.1 SyntaxError

Syntax errors occur when syntax requirements are not met, and are detected in the interpreting process before the program is executed. Syntax errors are removed when translating the source code into a binary code. When a syntax error is detected, the Python interpreter outputs the message "SyntaxError: invalid syntax." Such errors occur frequently, especially when you are unfamiliar with Python's syntax. Code 7.1 shows a syntax error with a missing colon (':') after True.

Unfortunately, error messages are often not informative. Described below are four common mistakes that result in syntax errors; this list can be used to help you detect problems:

- Indent correctly? In Code 7.2, *elif* statement should be at the same indentation level as *if* statement.
- Using the keywords correctly? Are there typos in the keywords?
- Using keywords as the names of variables, functions, or classes?
- Incomplete statements, such as
  - An unclosed bracket '{', ' (', '[', quote ', "
  - Omitting the colon symbol ":". For example, omitting the colon at the end of a *def* or *while* statement would yield a *SyntaxError: invalid syntax* message (Code 7.3).
  - Using if, elif, or while without any conditional expression.

To handle syntax errors, first check those four aspects.

```
>>> while True print 'hello world'
SyntaxError: invalid syntax
>>>
```

**CODE 7.1**
While invalid syntax problem.

```
>>> if (i>0):
        print 'i is bigger than 0'
        elif: print 'i is smaller than 0'

SyntaxError: invalid syntax
>>>
```

**CODE 7.2**
If invalid syntax because of indentation.

```
KeyboardInterrupt
>>> while True
SyntaxError: invalid syntax
>>> while True:
```
                                                        (a)
```
>>> def add()
SyntaxError: invalid syntax
>>> def add():
```
                                                        (b)

**CODE 7.3**
Missing parts of a statement syntax error.

### 7.1.3.2 Run-Time Exceptions

Errors detected during execution are called exceptions. If an exception is not fixed, the program is terminated and yields a so-called *traceback* error message (Code 7.4). Many operations could result in exceptions. For example, transferring a nondigit string to a float (Code 7.4a and b), accessing elements with index out of boundary (Code 7.4c), and numbers divided by 0 (Code 7.4d). Table 7.1 gives common exceptions and their corresponding causes.

```
>>> float([1,2,3])

Traceback (most recent call last):
  File "<pyshell#11>", line 1, in <module>
    float([1,2,3])
TypeError: float() argument must be a string or a number
```
                                                        (a)
```
>>> float('strv1')

Traceback (most recent call last):
  File "<pyshell#16>", line 1, in <module>
    float('strv1')
ValueError: could not convert string to float: strv1
```
                                                        (b)
```
>>> x = [1,2,3,4]
>>> x[4]

Traceback (most recent call last):
  File "<pyshell#14>", line 1, in <module>
    x[4]
IndexError: list index out of range
```
                                                        (c)
```
>>> 1/0
Traceback (most recent call last):
  File "<pyshell#18>", line 1, in <module>
    1/0
ZeroDivisionError: integer division or modulo by zero
```
                                                        (d)

**CODE 7.4**
Examples of run-time exceptions.

**TABLE 7.1**

Built-In Exceptions

| Class Name | Reasons for Having Exceptions |
| --- | --- |
| Exception | The root class for all exceptions |
| AttributeError | Attempt to access an undefined object attribute |
| IOError | Attempt to open a nonexistent file |
| IndexError | Request for a nonexistent index of a sequence, for example, list |
| KeyError | Request for a nonexistent dictionary key |
| NameError | Attempt to access an undeclared variable |
| SyntaxError | Code is ill-formed |
| TypeError | Pass function an argument with wrong type object |
| ValueError | Pass function an argument with correct type object but with an inappropriate value |
| ZeroDivionError | Division (/) or modulo (%) by a numeric zero |

Troubleshooting those errors are very specific when an exception is thrown during run time. The following steps can be taken to handle run-time exceptions:

- Check the exception type and review reasons causing the exceptions (Table 7.1).
- Look into the code, especially the line (indicated in the exception message) that throws errors, and analyze what resulted in the exception. Sometimes, you may need to go up/down a few lines to identify the real problem.
- If still not sure about the causes, use 'print' to output the values for relative variables to check if they are right.
- Revise code and run again.

### 7.1.3.3 Handling Exceptions

Handling Exceptions is the process of detecting the potential exceptions and dealing with them before the entire program fails (Python 2001b). *Try....* *except* statement is used to catch exceptions (Figure 7.7). Put the code that may produce run-time exceptions within the "try" block followed with the "except" block. There are three ways to handle the "except" part based on how specific the error message is following an exception.

You can catch each exception using a different except block (Figure 7.7a) or catch multiple exceptions within one except block (Figure 7.7b). You can also detect all exceptions without any exception type after except

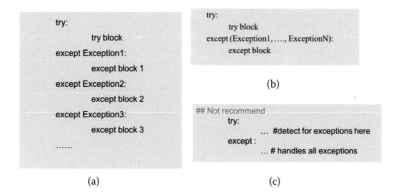

(a)

(b)

(c)

Heh.... No error any more.

*Hah...we have a list only having one item '\n', and therefore has no index '1'*

**FIGURE 7.7**
Raise and catch exceptions.

```
>>> try:
        pass #try block
>>> except:
        pass #except block
>>> finally:
        pass #finally block    #executes regardless    of exceptions
```

**CODE 7.5**
Uses "try...except...finally" to capture the ZeroDivisionError exception and remove result from memory if an exception happened.

(Figure 7.7c). Although the first approach requires more coding, it is recommended because it provides the most detailed exception messages for debugging. The last approach is not recommended because it does not provide potential causes for the exceptions.

The *try...finally* statement is used to define clean-up actions that must be executed under all circumstances (Code 7.5 and Code 7.6). This means that the *finally* statements are executed regardless of whether or not the exception occurs.

### 7.1.3.4 Add Exception Handles and Clean-Up Actions to File Read/Write

File operations may generate run-time exceptions, for example, the file or file path does not exist. A *try...except...finally...* statement can be applied to catch and handle the error. In Code 7.7, the file operation code is placed in the "try" block, "except" is used to capture IOError exceptions, and "finally" is used to close the file if it does exist.

```
>>> def divide(x,y):
        result = None
        try:
                result = x/y
                return result
        except ZeroDivisionError:
                print 'Division by zero'
        finally:
                print 'Cleaning up ....'
                del result


>>> divide(3,1)
Cleaning up ....
3
>>> divide(3,0)
Division by zero
Cleaning up ....
>>>
```

**CODE 7.6**
Try...except...finally to clean up variables if an exception occurred.

```
>>> f = None
>>> try:
        f = open('sample.txt', 'r+')
        f.readline()
        f.readlines()
        f.seek(0)
        f.read()
        f.write('This is a test!')
>>> except IOError:
        print 'The file does not exist!'
>>> finally:
        ## Close the file if the file opened
        if f:
            f.close()
```

**CODE 7.7**
Uses *try...except...finally* to handle file operation exceptions.

## 7.2 Python Modules

Modules are units for managing a complex set of codes within Python. Python modules are a convenient way for organizing, sharing, and reusing code. The essential knowledge for using modules includes (a) understanding the concept, (b) becoming familiar with common and useful Python modules, (c) managing codes as modules, and (d) configuring/installing a new module to the Python environment.

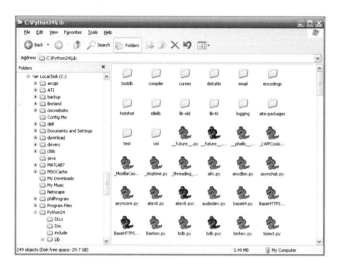

**FIGURE 7.8**
System modules.

### 7.2.1 Module Introduction

The code blocks that we developed during the hands-on practices have been saved as files (.py). Modules provide a logical way to organize and reuse these .py files. The Python code files can be imported as standard modules, such as sys and math module. Once imported, functions and attributes contained in the module can be used anywhere in the program.

A module is 'compiled' into a binary version .pyc. This process is automatically accomplished when the module is first loaded and executed. This compiling mechanism makes the module load faster but not run faster. Figure 7.8 shows the system modules, including .py, .pyc files, and packages.

During programming, it is possible to organize different classes, functions, and other components into different files to form different modules.

### 7.2.2 Set Up Modules

There are three ways to set up a path to allow the system to find the module. The easiest way is to update the system path temporarily by using the sys.path.append() method (Code 7.8). However, the path configuration is lost when the Python GUI is restarted.

To make the path configuration persistent, change the system's Environment Variable in the following steps:

1. Right click on "My computer" and select "Properties."
2. Click Advanced tab on the top; Click Environment Variables on the bottom.
3. Click New in either window.

```
>>> import sys
>>> sys.path.append('c:\ggs650code')
>>> sys.path
['C:/Users/Phil/Documents/courses/book - gis programming/codes',
 'C:\\WINDOWS\\System32', 'C:\\Python27\\ArcGIS10.2\\Lib\\idlelib',
 'C:\\WINDOWS\\SYSTEM32\\python27.zip', 'C:\\Python27\\ArcGIS10.2\\DLLs',
 'C:\\Python27\\ArcGIS10.2\\lib', 'C:\\Python27\\ArcGIS10.2\\lib\\plat-win',
 'C:\\Python27\\ArcGIS10.2\\lib\\lib-tk', 'C:\\Python27\\ArcGIS10.2',
 'C:\\Python27\\ArcGIS10.2\\lib\\site-packages',
 'C:\\Program Files (x86)\\ArcGIS\\Desktop10.2\\bin',
 'C:\\Program Files (x86)\\ArcGIS\\Desktop10.2\\arcpy',
 'C:\\Program Files (x86)\\ArcGIS\\Desktop10.2\\ArcToolbox\\Scripts',
 'c:\\ggs650code']
>>>
```

**CODE 7.8**
Add the path to the module through sys.path.append() method will add the path to the end of sys.path list.

4. Add the "PYTHONPATH" variable, and set the value as "c:\pythoncode."

5. Save and quit.

6. Restart your Python interactive GUI.

7. Check if your path can be found with sys.path by importing sys module.

Another method of setting up the module file path is to specify it at the beginning of the Python program as follows:

```
import sys, os
if os.getcwd() not in sys.path:
    sys.path.append(os.getcwd)
```

The first statement imports sys and os modules. The second statement fetches the current file path using os.getcwd to check if it is already in the system path, adding to the system path if not.

### 7.2.3 System Built-In Modules

Python comes with many built-in modules (Table 7.2). The popular ones include os, sys, math, shutil, and others. One rule for using a module is to import it first.

- The os module enables interactions with the operation system. For example, checking the current operation system, the file system, or determining the number of hard drives.
- The *sys* module provides system-specific parameters and functions. Common utility scripts often need to process command-line arguments. These arguments are stored in the sys module's argv attribute as a list.

**TABLE 7.2**

System Built-In Modules

| Module | Description | Examples |
|---|---|---|
| os | Interact with the operating system | os.system('dir') |
| sys | System-specific parameters and functions | sys.path, sys.exit() |
| math | Floating point math functions | math.pow() |
| shutil | High-level file operations | shutil.copyfile(), shutil.move() |
| glob | Get file lists from directory wildcard searches | glob.glob('*.shp') |
| re | Regular expression tools for advanced string processing | re.match('c', 'abcdef') |
| datetime | Manipulate dates and times | date.today() |
| zlib, gzip, bz2, zipfile and tarfile | Data archiving and compression | ZipFile('spam .zip', 'w') |

- *shutil* is a module providing high-level file operations.
- *glob* gets file lists from directory wildcard searches.
- *re* is a short module that provides regular expression tools for advanced string processing.
- The *random* module provides tools for making random selections.
- The *datetime* module supplies classes for manipulating dates and times in both simple and complex ways.
- Common data archiving and compression formats are directly supported by additional modules, including *zlib, gzip, bz2, zipfile,* and *tarfile*.

*math* (Figure 7.9) is the most popular module used in this book. We can always use built-in function dir() to find what is supported in a new module. Figure 7.10 shows that the math module includes many methods, such as acos, sin, fabs, etc., and several private attributes, such as '__doc__', '__name__', and '__package__'. These three attributes are typically included in all modules. The built-in function help() can be used to check the description of the module and functions.

## 7.3 Package Management and Mini-GIS

### 7.3.1 Regular GIS Data Organization

Figure 7.11 shows the ArcMap user interface managing multiple datasets. Each dataset is a layer after it is loaded from the disk and rendered on the

| | | |
|---|---|---|
| • degrees<br>• radians | >>> math.degrees(1)<br>57.295779513082323 | >>> math.radians(57.295)<br>0.99998639493015118 |
| • 'asin'<br>• 'sin' | >>> math.asin(0.5)<br>0.52359877559829893 | >>> math.sin(0.5)<br>0.479425538604203 |
| • 'cos'<br>• 'acos' | | |
| • 'exp'<br>• 'fabs' | >>> math.exp(2)<br>7.3890560989306504 | >>> math.fabs(-2)<br>2.0 |
| • 'log10'<br>• 'pi' | >>> math.log10(100)<br>2.0 | >>> math.pi<br>3.1415926535897931 |
| • 'pow'<br>• 'sqrt' | >>> math.pow(3, 2)<br>9.0 | >>> math.sqrt(9)<br>3.0 |

**FIGURE 7.9**
math methods.

```
Python Shell
File Edit Shell Debug Options Windows Help
>>> import math
>>> dir(math)
['__doc__', '__name__', '__package__', 'acos', 'acosh', 'asin', 'asinh', 'atan', 'atan2', 'atanh', 'ceil', 'copysign', 'cos', 'cosh', 'd
egrees', 'e', 'erf', 'erfc', 'exp', 'expm1', 'fabs', 'factorial', 'floor', 'fmod', 'frexp', 'fsum', 'gamma', 'hypot', 'isfinite', 'isinf
', 'isnan', 'ldexp', 'lgamma', 'log', 'log10', 'log1p', 'modf', 'pi', 'pow', 'radians', 'sin', 'sinh', 'sqrt', 'tan', 'tanh', 'trunc']
>>> help('math')
Help on built-in module math:

NAME
    math

DESCRIPTION
    This module is always available.  It provides access to the
    mathematical functions defined by the C standard.

FUNCTIONS
    acos(...)
        acos(x)

        Return the arc cosine (measured in radians) of x.

    acosh(...)
        acosh(x)

        Return the hyperbolic arc cosine (measured in radians) of x.

    asin(...)
        asin(x)

        Return the arc sine (measured in radians) of x.
```

**FIGURE 7.10**
Check a module.

map display area (Figure 7.12). The composite layers of a map show the street, park, house (Parcels), and river information about the city of Westerville, Ohio. Upon loading a dataset to the memory, each layer includes a series of objects (e.g., each house can be treated as one object).

### 7.3.2 Mini-GIS Package

In Chapters 1 through 6, we have developed a series of classes and functions (Figure 7.13). In our previous practice, the visualization module has been independent of the Point, Polyline, and Polygon module (Figure 7.13a). However, visualization is added as a method to the Point, Polyline, and Polygon module through the inheritance from a common super class, the

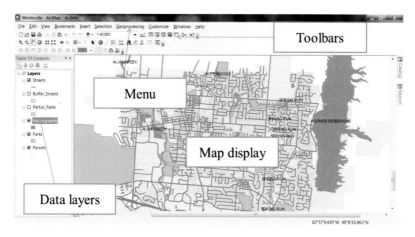

**FIGURE 7.11**
ArcMap user interface.

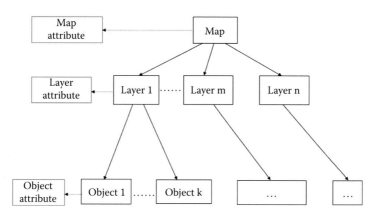

**FIGURE 7.12**
Map data organization hierarchy.

Feature class (Figure 7.13b). Therefore, the visualization does not have to be defined many times.

Putting these together creates a package for reading and displaying ESRI shapefiles, including simple functions (e.g., calculate the distance and centroid). As shown in Figure 7.14, Tkinter is a built-in module for data visualization in a GUI, and struct is used for reading binary data. A feature class is a simple module including only two methods, __init__() and vis() (Figure 7.14). Polyline and Point classes are inherited from the Feature class and include two methods, __init__() and vis(), as well. In addition to these two methods, Polylines also include a method length(). A polygon is inherited from the Polyline class and overrides the vis() function.

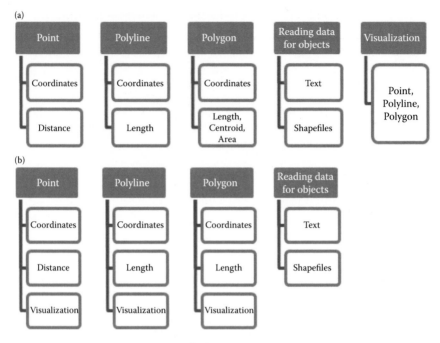

**FIGURE 7.13**
Developed code. (a) Visualization explicitly defined separate from Point, Polyline, and Polygon. (b) Visualization is inherited from Feature class therefore, no need to define visualization explicitly.

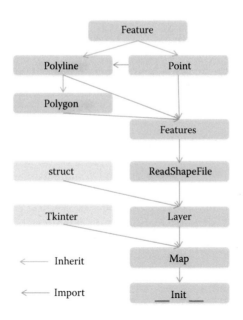

**FIGURE 7.14**
Hierarchy of primary Mini-GIS modules (Tkinter and struct are Python built-in modules).

## 7.4 Hands-On Experience with Mini-GIS

Mini-GIS is a tiny GIS software package developed using Python. The source code can be downloaded from the course material package. After downloading and unzipping the miniGIS.zip file, there will be a miniGIS folder containing several Python files and a sample data folder shown in Figure 7.15.

In the folder, *.py files are the Python files and *.pyc files are the compiled python files. The 'sample-data' folder contains several shapefile data folders; the 'cities' folder, which contains world city data; 'countries,' which contains world country data; the 'Fairfax' folder contains schools, highways, major utility lines, and regions in Fairfax; and the 'multi-parts-polyline' folder contains two polylines with multiparts.

### 7.4.1 Package Management and Mini-GIS

The ReadShapeFile module reads different types of features, or layers, which include Points, Polylines, and Polygons (Figure 7.16). The Commons

| Name | Date modified | Type | Size |
|------|---------------|------|------|
| sample-data | 9/25/2014 11:32 AM | File folder | |
| Commons | 9/25/2014 10:54 AM | Compiled Python ... | 3 KB |
| Dialogs | 9/25/2014 10:54 AM | Compiled Python ... | 3 KB |
| Events | 9/25/2014 10:54 AM | Compiled Python ... | 7 KB |
| Feature | 9/25/2014 10:56 AM | Compiled Python ... | 2 KB |
| Layer | 9/25/2014 11:13 AM | Compiled Python ... | 3 KB |
| LineSegment | 9/25/2014 10:56 AM | Compiled Python ... | 3 KB |
| Map | 9/25/2014 11:36 AM | Compiled Python ... | 8 KB |
| Point | 9/25/2014 10:56 AM | Compiled Python ... | 2 KB |
| Polygon | 9/25/2014 10:56 AM | Compiled Python ... | 2 KB |
| Polyline | 9/25/2014 10:56 AM | Compiled Python ... | 4 KB |
| ReadShapeFile | 9/25/2014 10:56 AM | Compiled Python ... | 3 KB |
| Commons | 9/25/2014 1:21 AM | Python File | 2 KB |
| Dialogs | 9/25/2014 1:22 AM | Python File | 3 KB |
| Events | 9/25/2014 1:24 AM | Python File | 5 KB |
| Feature | 9/24/2014 12:00 PM | Python File | 1 KB |
| Layer | 9/25/2014 11:13 AM | Python File | 2 KB |
| LineSegment | 9/23/2014 8:17 PM | Python File | 4 KB |
| Map | 9/25/2014 11:36 AM | Python File | 8 KB |
| MiniGIS | 9/25/2014 11:40 AM | Python File | 4 KB |
| Point | 9/24/2014 12:29 PM | Python File | 1 KB |
| Polygon | 9/25/2014 1:27 AM | Python File | 1 KB |
| Polyline | 9/25/2014 1:30 AM | Python File | 4 KB |
| ReadShapeFile | 9/25/2014 10:56 AM | Python File | 3 KB |

**FIGURE 7.15**
Mini-GIS code organization.

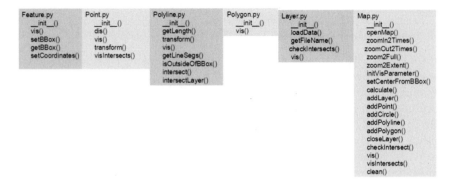

**FIGURE 7.16**
Primary functions are feature modules and map modules.

module provides common reusable functions; The Dialogs Module provides the two dialogs to select layers; and the Events module provides mouse-operated events handling.

*Mini-GIS* module includes the main function. A directory becomes a Python package if it contains __init__.py. It includes the following functions:

1. The main functions for calling other parts
2. Generate a window
3. Input the datasets
4. Visualize the data
5. Show the window

Normally, this is the only part you need to understand fully if you get the package from elsewhere; otherwise, the interface description is enough to use a package.

### 7.4.2 Run and Practice the Mini-GIS Package

We can run the package in the following workflow:

1. Add the path to the sys path; or just double click the MiniGIS.py; or use IDLE to run the package. Here, use the third method to open MiniGIS.py.
2. Click Run→Run module menu or the 'F5' button to run the package. It will bring up a window, which has the title 'MiniGIS.'
3. Import shapefile in two ways: 'Import shp' and 'Add shp layer' in the File menu. Figure 7.17a illustrates the map of imported Fairfax shapefile folder data.

(a) (b)

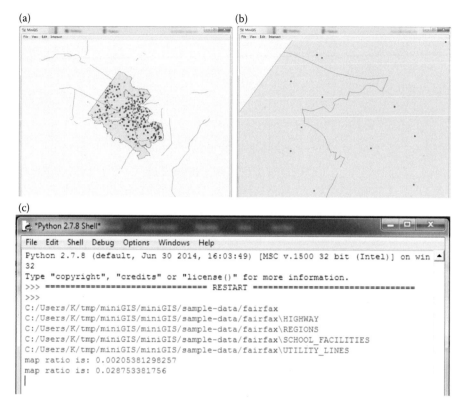

(c)

**FIGURE 7.17**
(a) Using Mini-GIS to import Fairfax shapefile folder. (b) Zoom to extent. (c) Python Shell outputs.

4. View the map using the menu under 'View' menu: zoom in, zoom out, zoom to extent, zoom to full window and close layer. Figure 7.17b shows zooming the map to the left boundary. Figure 7.17c shows the outputs in the Python Shell, which also outputs the map ratio.

Mini-GIS also supports drawing features of different types: point, polyline, circle, and polygon. To draw a feature, click on 'Edit' to bring up a dialog box and select a layer. Figure 7.18 shows the feature added to the map in the bottom left corner. However, the current Mini-GIS does not support exporting these added features to the shapefile.

The Mini-GIS can also calculate the intersections between two polyline layers. Figure 7.19 indicates the intersections between utility layers and highway layers with yellow circles.

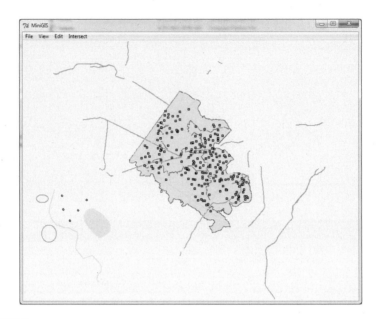

**FIGURE 7.18**
Draw features.

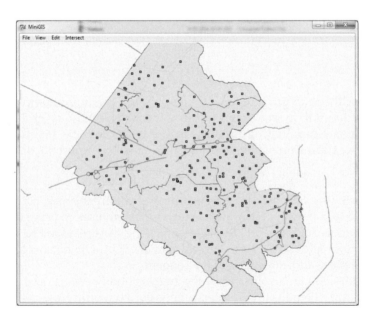

**FIGURE 7.19**
Intersections between HIGHWAY and UTILITY_LINES layers.

## 7.5 Chapter Summary

The chapter reviews the IDE of Python and demonstrates how to customize the settings of a Python IDE. Modules are also discussed and a Mini-GIS package is introduced for reading and visualizing ESRI Shapefiles. The implementation details (such as reading from shapefiles) of the Mini-GIS will be introduced in the following chapters.

### PROBLEMS

1. Take the data used in Chapter 4.
2. Based on the solution to problems in Chapter 4, how can you find the bounding box (minx, miny, maxx, maxy) of the polylines?
3. Prepare the data to visualize (coordinates transform, etc.) based on the window size you want to draw.
4. Program to draw the data (using Canvas and related objects/functions/widgets). Organize your program in a file or several files to set up a module in the system; import the module and execute the module that was developed and configured.
5. Develop a word document explaining the program.
6. Provide comments for all lines in the code to explain the logic.

# 8

## Vector Data Algorithms

GIS is different from other software technologies due to its algorithms and analyses for spatial relationships in spatial data. This chapter introduces a similar category of algorithms for vector data processing, including the calculations of centroid (8.1), area (8.2), length (8.3), line intersection (8.4), and point in polygon (8.5), which are the fundamental algorithms based on geometry and the spatial relationship.

## 8.1 Centroid

Centroid can be considered as the gravity center of a feature (wiki.gis 2011). One application example of centroid is to set up the annotation location, such as labeling a building on a map. This section discusses the centroid calculation for three geometry types: triangle, rectangle, and polygon.

### 8.1.1 Centroid of a Triangle

A triangle's centroid is defined as the intersection point of three medians of a triangle (Johnson 1929, Figure 8.1). A triangle's median is the line segment extending from one vertex to the midpoint of its opposite side. Given three vertices $(x_1,y_1)$, $(x_2,y_2)$, and $(x_3,y_3)$ of a triangle, the centroid $(x_{centroid}, y_{centroid})$ can be calculated with Equation 8.1.

$$\begin{cases} x_{centroid} = \dfrac{(x_1 + x_2 + x_3)}{3} \\ y_{centroid} = \dfrac{(x_1 + x_2 + x_3)}{3} \end{cases} \tag{8.1}$$

### 8.1.2 Centroid of a Rectangle

A rectangle's centroid is the intersection of the rectangle's two diagonals (Figure 8.2). Given the bottom-left vertex $(x_{min}, y_{min})$ and upper-right

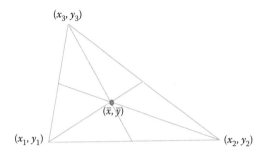

**FIGURE 8.1**

A triangle's centroid is the intersection point of the three medians. (Refer to http://jwilson.coe.uga.edu/EMAT6680Su09/Park/As4dspark/As4dspark.html. for the proof of concurrency of the three medians.)

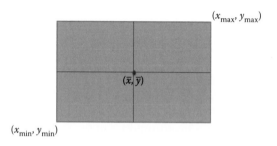

**FIGURE 8.2**

A rectangle's centroid is the intersection of two diagonals.

vertex $(x_{max}, y_{max})$ of a rectangle, the centroid can be calculated with Equation 8.2.

$$\begin{cases} x_{centroid} = \dfrac{(x_{min} + x_{max})}{2} \\ y_{centroid} = \dfrac{(y_{min} + y_{max})}{2} \end{cases} \tag{8.2}$$

### 8.1.3 Centroid of a Polygon

For any non-self-intersecting closed polygon with $n$ points $(x_0, y_0)$, $(x_1, y_1)$, ..., $(x_{n-1}, y_{n-1})$, the centroid can be calculated with Equation 8.3, where $A$ is the area of the polygon, which is discussed in Section 8.2 (Bourke 1988). In this formula, the last vertex $(x_n, y_n)$ of the polygon is assumed to be the same as the first vertex $(x_0, y_0)$. In fact, triangles and rectangles are special polygons, so Equation 8.3 can also be used for calculating the centroid of triangle and rectangle.

$$\begin{cases} x_{\text{centroid}} = \dfrac{\displaystyle\sum_{i=0}^{n-1}(x_{i+1}+x_i)(x_iy_{i+1}-x_{i+1}y_i)}{6A} \\[4ex] y_{\text{centroid}} = \dfrac{\displaystyle\sum_{i=0}^{n-1}(y_{i+1}+y_i)(x_iy_{i+1}-x_{i+1}y_i)}{6A} \end{cases} \tag{8.3}$$

## 8.2 Area

This section introduces how to calculate the area of polygon feature. Two types of polygons are discussed: a simple polygon and a polygon with hole(s).

### 8.2.1 Area of a Simple Polygon

A simple polygon is a polygon without self-intersecting sides that forms a closed path. Figure 8.3a shows an example of a simple polygon with six points (vertices), where point 1 is the same as point 6, forming a closed path. The points are recorded in clockwise order.

To calculate the area of this polygon, we first draw vertical lines from each point of the polygon to the x-axis to form five trapezoids: A and B (Figure 8.3b), and C, D, and E (Figure 8.3c). By visually examining the five trapezoids, the area of the polygon ($S_P$) can be derived from Equation 8.4.

$$S_P = S_A + S_B - S_C - S_D - S_E \tag{8.4}$$

where $S_A$, $S_B$, $S_C$, $S_D$, and $S_E$ denote the corresponding areas of the five trapezoids.

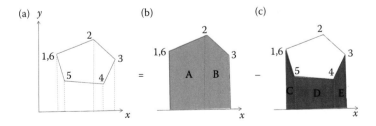

**FIGURE 8.3**
Area calculation of a simple polygon.

Since the coordinates for each point are known, we can calculate the area for each trapezoid with Equation 8.5.

$$S_A = \frac{(x_2 - x_1)(y_2 + y_1)}{2}$$

$$S_B = \frac{(x_3 - x_2)(y_3 + y_2)}{2}$$

$$S_C = \frac{(x_4 - x_3)(y_4 + y_3)}{2} \tag{8.5}$$

$$S_D = \frac{(x_5 - x_4)(y_5 + y_4)}{2}$$

$$S_E = \frac{(x_6 - x_5)(y_6 + y_5)}{2}$$

By plugging Equation 8.5 into Equation 8.4, we derive Equation 8.6 to calculate the area of the polygon.

$$S_P = \sum_{i=1}^{5} \frac{(x_{i+1} - x_i)(y_{i+1} + y_i)}{2} = \frac{1}{2} \sum_{i=1}^{5} (x_{i+1}y_i - x_i y_{i+1}) \tag{8.6}$$

By generalizing Equation 8.6, we have Equation 8.7 to calculate the area for any simple polygon with $n$ points (point$_0$ = point$_n$).

$$A = \frac{1}{2} \sum_{i=0}^{n-1} (x_i y_{i+1} - x_{i+1} y_i) \tag{8.7}$$

## 8.2.2 Area of a Polygon with Hole(s)

A polygon could have one or more holes. For instance, when a lake is represented as a polygon, an island in the lake can be represented as a hole in the polygon. When calculating area for a polygon, we need to subtract the area(s) of the hole(s) from the area of the encompassing polygon. Figure 8.4 shows a

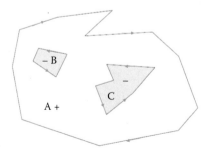

**FIGURE 8.4**
A polygon with two holes.

polygon (A) with two holes (B and C). In GIS, the vertices of a hole are recorded in counterclockwise order. Hence, the area of the holes $S_B$ and $S_C$ calculated using Equation 8.7 will result in a *negative number* ($S_A < 0$, $S_B < 0$). Therefore, the actual area of polygon A can be calculated as $S = S_A - (-S_B) - (-S_C) = S_A + S_B + S_C$.

## 8.3 Length

Length calculation finds the length of a line feature and the perimeter of a polygon feature.

### 8.3.1 Length of a Straight Line Segment

Suppose there are two points: $P_1(x_1, y_1)$ and $P_2(x_2, y_2)$ in the 2D Cartesian coordinate system. To calculate the length of the line segment $P_1P_2$, we draw a horizontal line from $P_1$ and a vertical line from $P_2$ to form a right triangle (Figure 8.5). Suppose that the length for the horizontal side is $a$, for the vertical side is $b$, and the hypotenuse is $c$. According to the Pythagorean theorem, we have the equation $c^2 = a^2 + b^2$.

If the side lengths of the two right angles $a$ and $b$ are known, the length of the third side, which is the length of the line segment $P_1P_2$, can be derived. Since $a$ and $b$ are parallel to the axis, we have $a = x_2 - x_1$, $b = y_2 - y_1$. Plugging these two equations to $c^2 = a^2 + b^2$, Equation 8.8 is derived to calculate the length of a line segment:

$$P_1P_2 = c = \sqrt{(x_2 - x_1)^2 + (y_2 - y_1)^2} \tag{8.8}$$

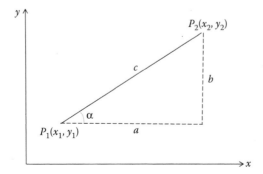

**FIGURE 8.5**
Length calculation of a straight line segment.

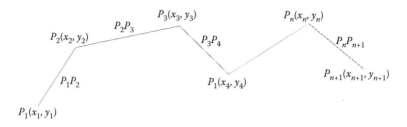

**FIGURE 8.6**
A polyline with $n + 1$ points.

## 8.3.2 Length of a Polyline

A polyline consists of many line segments, so the length of a polyline is the sum of the length of all line segments.

Suppose there is a polyline with $n + 1$ points: $p_1, p_2, p_3, \ldots, p_n$, and $p_{n+1}$ (Figure 8.6). Based on Equation 8.8, the length of this polyline can be calculated by summing the lengths of all line segments (Equation 8.9).

$$L_{\text{polyline}} = \sum_{i=1}^{n} P_i P_{i+1} = \sum_{i=1}^{n} \sqrt{(x_{i+1} - x_i)^2 + (y_{i+1} - y_i)^2} \qquad (8.9)$$

## 8.4 Line Intersection

Checking intersections between two lines is fundamental for the many spatial relationship calculations. When determining line intersections, we need to check whether or not the two line segments have shared point(s). Given two line segments, Line1 $((x_1,y_1), (x_2,y_2))$ and Line2 $((x_3,y_3), (x_4,y_4))$ in the Cartesian coordinate system (Figure 8.7), if two line segments intersect then there must be at least one point $(x_0,y_0)$ that is on both segments. So the question is how to find out whether there is a common point for the two line segments. Furthermore, how do we calculate the coordinate values of the common point $(x_0,y_0)$?

A common method to represent a line in the Cartesian coordinate system is to use the linear equation: $y = ax + b$, where $x$, $y$ are the variables, $a$, $b$ are the constants or parameters, $a$ is the slope of the line, and $b$ is the intercept of the line with $y$-axis (Figure 8.8).

Based on this, Line1 can be represented as $y = a_{12}x + b_{12}$; and Line2 can be represented as $y = a_{34}x + b_{34}$, where $a_{12}, b_{12}, a_{34}, b_{34}$ are constant values that can be calculated based on the four points $(x_1,y_1), (x_2,y_2), (x_3,y_3)$, and $(x_4,y_4)$ as detailed below.

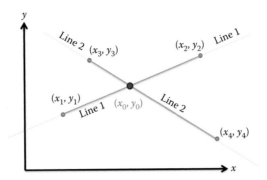

**FIGURE 8.7**
Illustration of line segments intersection.

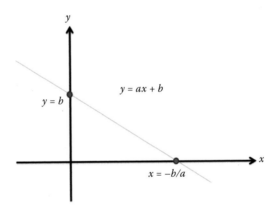

**FIGURE 8.8**
Mathematical representation of a line in Cartesian coordinate system.

*Calculating $a_{12}$, $b_{12}$, $a_{34}$, $b_{34}$:* Since the two points $(x_1, y_1)$, $(x_2, y_2)$ are on Line1, we have the following equation group (Equation 8.10):

$$\begin{cases} y_1 = a_{12}x_1 + b_{12} \\ y_2 = a_{12}x_2 + b_{12} \end{cases} \tag{8.10}$$

By solving Equation 8.10, $a_{12}$, $b_{12}$ is represented as Equation 8.11:

$$\begin{cases} a_{12} = \dfrac{y_2 - y_1}{x_2 - x_1} \\ b_{12} = y_1 - a_{12}x_1 \end{cases} \tag{8.11}$$

where $x_2$ is not equal to $x_1$.

Similarly, for Line2, we have the following equation group (Equation 8.12):

$$\begin{cases} y_3 = a_{34}x_3 + b_{34} \\ y_4 = a_{34}x_4 + b_{34} \end{cases} \tag{8.12}$$

By solving Equation 8.12, $a_{34}$, $b_{34}$ can be represented as Equation 8.13:

$$\begin{cases} a_{34} = \dfrac{y_4 - y_3}{x_4 - x_3} \\ b_{34} = y_3 - a_{34}x_3 \end{cases} \tag{8.13}$$

where $x_3$ is not equal to $x_4$.

As mentioned above, if Line1 and Line2 intersect, there must be a point $(x_0, y_0)$ that is common to both line segments. In such case, $(x_0, y_0)$ is the solution for the binary linear equation group (Equation 8.14):

$$\begin{cases} y = a_{12}x + b_{12} \\ y = a_{12}x + b_{12} \end{cases} \tag{8.14}$$

Accordingly, the problem of looking for the common point is transformed to checking whether or not the equation group 8.14 has a solution. The solution (if exists) for Equation 8.14 can be represented as Equation 8.15:

$$\begin{cases} x_0 = \dfrac{b_{12} - b_{34}}{a_{34} - a_{12}} \\ y = a_{12}x_0 + b_{12} \end{cases} \tag{8.15}$$

where $a_{34}$ is not equal to $a_{12}$, and $a_{12}$, $b_{12}$, $a_{34}$, $b_{34}$ can be calculated based on the given four points. It should be noted that $(x_0, y_0)$ is the solution for two infinite lines. To check whether $(x_0, y_0)$ falls on both Line1 $((x_1, y_1), (x_2, y_2))$ and Line2 $((x_3, y_3), (x_4, y_4))$, the following test is required:

$$x_1 \le x_0 \le x_2$$
$$x_3 \le x_0 \le x_4$$
$$y_1 \le y_0 \le y_2$$
$$y_3 \le y_0 \le y_4$$

If the four conditions are all true, Line1 and Line2 intersect at point $(x_0, y_0)$. Otherwise, they do not intersect.

The above method for checking the intersection does not work for two special scenarios: parallel lines and vertical lines.

### 8.4.1 Parallel Lines

Since parallel lines have the same slope, $a_{34}$ is equal to $a_{12}$. Therefore, Formula 8.15 does not work for parallel lines (divided-by-zero when $a_{34} = a_{12}$). This makes sense since parallel lines will never intersect. However, if two line segments are on the same line, we also have $a_{34} = a_{12}$. For this scenario, the two line segments will either overlap (have one or more than one shared point) or not intersect.

### 8.4.2 Vertical Lines

As illustrated in Figure 8.9, if Line1 $((x_1,y_1), (x_2,y_2))$ is vertical, we have $x_2 = x_1$. In this case, the $x$ coordinate of the intersection point is equal to $x_1$ $(x_0 = x_1)$, and the $y$ coordinate of the intersection point $(y_0)$ can be calculated as by plugging $x_0$ to the equation of Line2 as Equation 8.16.

$$y_0 = a_{34}x_0 + b_{34} \qquad (8.16)$$

Once $(x_0,y_0)$ is calculated, the same test is required to check whether the intersection point falls on both line segments. The same method can be applied if Line2 $((x_3,y_3), (x_4,y_4))$ is vertical.

If both lines are vertical, they are parallel. This can be handled the same way as parallel lines.

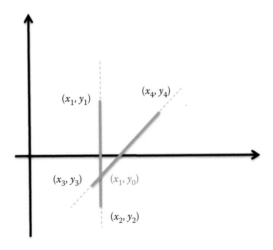

**FIGURE 8.9**

Line1 $((x_1,y_1), (x_2,y_2))$ is vertical.

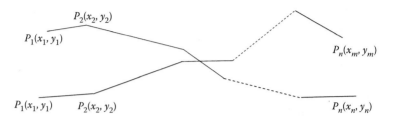

**FIGURE 8.10**
Intersection checking of two polylines.

Since a polyline consists of two or more line segments (Figure 8.10), the following procedure can be used to check whether the two polylines intersect: for each line segment in polyline1, check whether it intersects with any of the line segments in polyline2. Once a single intersection point is found, we can conclude that the two polylines intersect. We may also find out all the intersection points of two polylines by checking each of the line segment pairs between the two polylines.

## 8.5 Point in Polygon

A ray casting algorithm (Shimrat 1962) is often used to check whether a point is inside a polygon. With a ray casting algorithm, a ray is first drawn starting from the point to any fixed direction (normally take a vertical or horizontal direction to simplify the calculation), and then test how many intersection points that ray has with the polygon. If the number of the intersection points is odd, the point is in the polygon, otherwise, the point falls outside of the polygon.

Figure 8.11 illustrates how the ray casting works for testing whether or not each of the seven given points are in the polygon. Table 8.1 summarizes the number of intersection points for the seven vertical rays.

### 8.5.1 A Special Scenario

If the ray passes the polygon vertex, then the ray intersects with both edges sharing the vertex. Counting this scenario as two intersection points is problematic. In Figure 8.12, the ray of point1 passes a vertex, which generates two intersection points, incorrectly suggesting that this point is outside of the polygon based on the odd–even rule. For point3, which is outside the polygon, the number of intersections is also two, which is

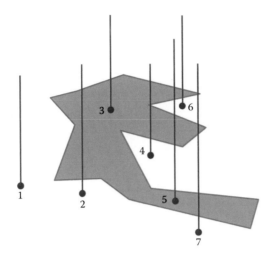

**FIGURE 8.11**
Illustration of ray casting algorithm.

**TABLE 8.1**

Number of Intersection Points for the Seven Rays

| Point Id | Number of Intersection Points | Point in Polygon? |
|----------|-------------------------------|-------------------|
| 1 | 0 | No |
| 2 | 2 | No |
| 3 | 1 (odd) | Yes |
| 4 | 4 | No |
| 5 | 5 (odd) | Yes |
| 6 | 2 | No |
| 7 | 6 | No |

consistent with the rule. Special consideration is needed to resolve such scenarios.

Examining carefully, the difference is that, for point1, the two edges sharing the vertex are on different sides of the ray, while for point three, the two edges are on the same side (left) of the ray. A general pattern is that if the edges are on different sides, we consider it as one intersection point, but if the edges are the same side, then we consider it as two intersection points. Based on this pattern, both edges of point2 are on the right side of the ray as it passes through the vertex (i.e., two intersections); adding the unambiguous intersection at the top of the polygon results in a total of three intersection points, and we can correctly conclude that it falls inside the polygon.

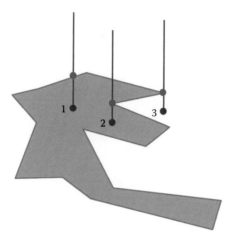

**FIGURE 8.12**
Special scenarios: ray passes the polygon vertex.

## 8.6 Hands-On Experience with Python

### 8.6.1 Using Python to Draw a Polygon and Calculate the Centroid

Download the Python file chap8_centroid_pract.py from the course material package and run the code to get the "Centroid Practice" window. Select
"Draw Polygon" and click on the canvas to draw the polygon, then click
"Finish Draw" to finish the drawing. Select "Calculate Centroid" to calculate
and visualize the centroid on the canvas. Identify and analyze the code for
the centroid calculation and visualization (Figure 8.13).

### 8.6.2 Using Python to Draw Polygon and Calculate the Area of Polygon

Download the Python file chap8_area_pract.py from the course material
package and run the code to bring up the "Area Practice" window. Select
"Draw Polygon" and click on the canvas to draw a polygon, then click on
"Finish Draw" to finish drawing the image. Finally, click on "Calculate
Area" to show the area of this polygon. Identify and analyze the code for the
area calculation (Figure 8.14).

### 8.6.3 Using Python to Draw Line Segments and Calculate the Intersection

Download the Python file chap8_intersection_pract.py from the course
material package and run the code to get the "Line Intersection Practice"
window. Select "Draw Lines" and click on the canvas to draw one line, then
click "Finish Draw". Repeat to draw another line, then click the "Check
Intersection" button to see whether they intersect. Please identify and
analyze the code for intersection calculation (Figure 8.15).

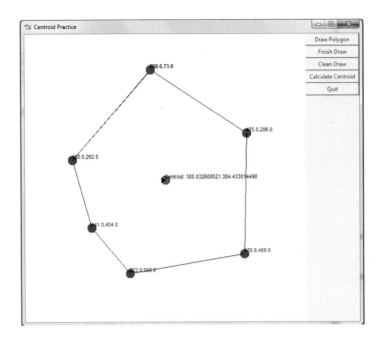

**FIGURE 8.13**
"Centroid Practice" window.

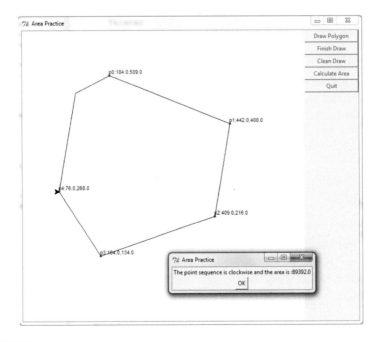

**FIGURE 8.14**
"Area Practice" window.

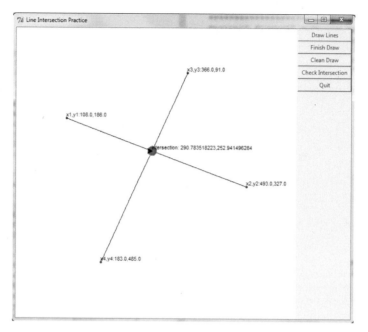

**FIGURE 8.15**
"Line Intersection Practice" window.

## 8.7 Chapter Summary

This chapter introduces several popular vector data algorithms within GIS, including line intersections, centroid calculations, area and length calculations, and point-in-polygons. Following the discussion of the algorithms, hands-on practices are provided, aiming to demonstrate how to programmatically implement the vector data algorithms using Python.

### PROBLEMS

- Review the class material.
- Review the class practice code: chap8_intersection_pract.py
- Create the following four line segments: [(50,200),(400,200)],[(60,450), (400,450)],[(100,600),(350,250)],[(300,100),(300,400)]
- Write a program to determine whether they intersect with each other.
- Display the line segments on the monitor and draw the intersected point in a different color and size.

# Section III

# Advanced GIS Algorithms and Their Programming in ArcGIS

# 9

## ArcGIS Programming

Programming all GIS analysis or mapping functions from scratch is not feasible because of their complexities; however, many commercial and open-source packages, such as ArcGIS, support such comprehensive GIS functions. Integrating various functions from existing packages will satisfy analytical requirements of an application. ArcGIS supports such workflow implementation through Python programming, which has become a sought-after capability for GIS professionals. This chapter introduces programming in ArcGIS for Desktop (i.e., ArcMap) using Python. The ArcGIS knowledge introduced in this chapter will set the foundation for later chapters when programming for algorithms and analyses.

### 9.1 ArcGIS Programming

Many practical GIS applications are complex and involve multiple GIS functions, such as modifying and querying features and their associated attributes. Taking hospitalization as a typical GIS example, suppose there are a group of potential patients who would like to go to the nearest hospitals. In order to locate the nearest hospital for each potential patient, iteration needs to go through every patient to identify the hospitals within that patient's neighborhood and then locate the nearest hospital of each potential patient. This becomes cumbersome when there are a large number of patients.

Manually conducting and repeating spatial data querying and processing is time-consuming, tedious, and error-prone. It is, therefore, beneficial to automate such processes using the functionality of a sophisticated GIS package. ArcGIS ModelBuilder (Figure 9.1, Esri 2016a) allows users to integrate multiple tools from ArcToolbox (Figure 9.1a) to form an analysis workflow in a graphical user interface. Note that ArcToolbox is reorganized in the new version of ArcGIS (e.g., 10.4 and pro version). The related information can be found at http://pro.arcgis.com/en/pro-app/help/analysis/geoprocessing/basics/what-is-geoprocessing-.htm. The model could batch the processes supported by the tools. Additionally, ArcGIS provides ArcPy package (Esri 2016b), a Python library/module that can operate most ArcMap functions. Automating a process involving multiple GIS algorithms or steps using the ArcPy package and Python is both fast and easy. The batch process built into ModelBuilder can also be exported as Python scripts. ModelBuilder is,

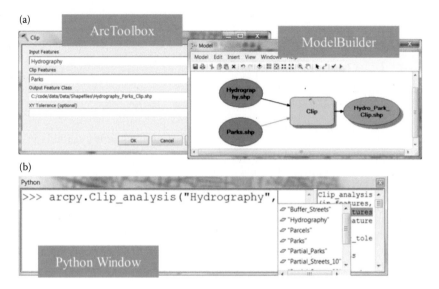

**FIGURE 9.1**
Geoprocessing with (a) ArcToolbox, ModelBuilder, and (b) ArcMap Python window.

therefore, similar to Python scripting with ArcPy; however, scripting with ArcPy is more flexible than ModelBuilder, which is restricted by available ArcTools and operation template. ArcMap provides a plug-in Python editing window (Figure 9.1b). You can write Python scripts in a Python IDE outside ArcMap by importing the ArcPy package. This chapter introduces how to use ArcPy and demonstrates its use through a number of examples. Section 9.2 details ArcPy structure, the programming environment, and how to use help documents. Sections 9.3 through 9.7 present several of the most frequently used ArcPy functionalities. Sections 9.8 and 9.9 demonstrate how to add the developed Python scripts as a new ArcTool so that others can reuse the script tool from ArcMap.

## 9.2 Introduction to ArcPy Package

### 9.2.1 ArcPy Functions, Classes, and Modules

ArcPy package is built on the previous 'ArcGIS geoprocessor scripting module' provided to users since ArcGIS 10.0 version to offer functions, classes, and modules for accessing most ArcMap functions:

- Function: An Esri-defined functionality that finishes a specific task (e.g., ListFeatureClasses, Describe, SearchCursor).

- Class: An Esri-defined object template that has a set of properties and methods (e.g., SpatialReference, FieldMap objects).
- Module: A Python file that generally includes functions and classes that finish similar tasks. There are data access, mapping, network analyst, spatial analyst, and time modules in ArcMap 10.1+ versions.

### 9.2.2 Programming with ArcPy in ArcMap

Beginning with version 10.0, Esri had embedded a Python window inside ArcMap (Figure 9.2 upper), which can be opened through the button on the toolbar (inside the red rectangle). We can type Python script in the left panel of the Python window (Figure 9.2 lower), which has ArcPy imported, so we do not have to import the package from scripting. The interactive help document is included in the right panel to describe the current function or class being used.

NOTE: We use 'ArcPy' in this text to be consistent with the Esri naming convention; however, in programming, 'arcpy' (all lowercase) will be used.

Hands-On Practice 9.1

- Open Python window from ArcMap and type in (not copy) the following command:

    *arcpy.Buffer_analysis("","","", "FULL", "FLAT", "ALL")*

**FIGURE 9.2**
ArcMap Python window and the opening button.

- While typing, observe the change in the interactive help document on the right panel of the window.

### 9.2.3 Programming with ArcPy in Python Window outside ArcMap

If programming using ArcPy outside ArcMap, import the ArcPy package (Figure 9.3) first, and then import the usage of ArcPy in the external environment to ensure that it is the same as in the ArcMap Python window. ArcGIS must be installed to use ArcPy. The scripts developed can be saved as *.py files, distributed to others, or added as a new ArcTool in the toolbox, as detailed in Section 9.8.

To improve the flexibility of the developed scripts for end users, configure the input as "dynamic" value, and allow the program to automatically retrieve the input. The *sys.argv* can be used for this purpose with a list of strings that hold the script name and additional arguments that are passed to the script.

Hands-On Practice 9.2: Fetching System Parameter Input

1. Open the Python editor window and enter Code 9.1. Save the code to a .py file, for example, samplecode.py.
2. Open the windows command console, navigate to the folder where the *.py file was saved, and run the script to perform buffer analysis with the following command:

   *samplecode.py   C:\Default.gdb\out2   C:\Default.gdb\output   5000 Meters*

The first parameter is the absolute path of the input data, the second parameter is the absolute path of the output data, and the third describes the buffer size (Figure 9.4).

```
>>> import arcpy
>>> fc = r'C:\Users\Min\Documents\ArcGIS\Default.gdb\test4'
>>> fields = arcpy.ListFields(fc)
>>> for field in fields:
        print field.name

OBJECTID
Shape
STATEFP
COUNTYFP
LINEARID
FULLNAME
RTTYP
MTFCC
Shape_Length
>>> |
```

**FIGURE 9.3**
Sample code of using ArcPy outside ArcMap.

```
import arcpy
import sys

script_name = sys.argv[0]
fc=sys.argv[1]
output=sys.argv[2]
bufferSize=sys.argv[3]
arcpy.Buffer_analysis(fc, output, bufferSize)
```

**CODE 9.1**
Python code using ArcPy and allowing flexible parameter inputs.

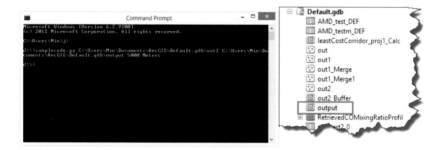

**FIGURE 9.4**
Screenshot of the windows command to execute a .py file and the output. (If there is no error reported in the window, and the output is generated, then the execution is successful.)

### 9.2.4 Using Help Documents

Esri offers various help documents for users to obtain detailed scripting knowledge when using ArcPy. Users can access the desktop help document (Figure 9.5) by clicking the "help" option on the ArcMap menu bar. Another help document is available on the ArcGIS resource center (Figure 9.5). For ArcGIS 10.1 to 10.2.2 version, the content is under "desktop" -> "geoprocessing" -> "ArcPy" at the resource center (http://resources.arcgis.com/en/help/main/10.2/index.html). For the most recent version, the information can be explored under http://resources.arcgis.com/en/help/. Both the resource center and the desktop version 'Help' document describe how to use ArcTools. At the end of each tool introduction, there are also syntax details and a sample code for using the ArcPy function (Figure 9.6). This can be accessed under "desktop" -> "geoprocessing" -> "tool reference." You can also search the name of the tool and the syntax so that the sample code is provided at the bottom of the results page.

For novice users, one way to learn ArcPy is to go through one of the three help documents, try the sample code, and make changes to the code using customized data. The Python prompt window inside ArcMap embeds the interactive help on each ArcPy function and class.

**FIGURE 9.5**
Desktop and online help document.

Hands-On Practice 9.3: Learn to Read Help Documents

1. Read the help in ArcGIS resource center.
2. Search "Add Field" tool in the search box on the top right.
3. Read through the syntax of "arcpy.AddField_management" and execute the sample code with your own data.

## 9.3  Automating ArcTools with Python

Most tools in ArcToolBox can be accessed using the ArcPy scripting. For example, the clip analysis is

**FIGURE 9.6**

Syntax and sample code of Python script in the online help document for the corresponding geoprocessing tool.

*arcpy.Clip_analysis (in_features, clip_features, out_feature_class, { cluster_tolerance})*

where the first three parameters are required and the last parameter with braces is optional. Scripting, rather than operating the tool through the ArcToolBox interface, is very useful when the process involves a loop (*for/ while statement*) or conditional expression (*if* statement) to execute the geoprocessing function repeatedly under certain conditions. For example, to calculate the line length within each polygon in a data layer, the process needs to include a *for* loop to enable the repeated process for each polygon. Inside the *for* loop, there will be a clip function that can cut off the line inside that polygon and a statistics function, which sums up the total length of the lines within the polygon.

Hands-On Practice 9.4: Calculate the Road Length within an Area

1. Type Code 9.2 in the Python window in ArcMap. Change the paths according to your workspace.

   Note that, arcpy.env.workspace is executed at the beginning to set the workspace, which is the path for accessing the input data and saving the results. With environment workspace set up, the input and output parameter can be set using the relative path. If the inputs are feature class, image, etc., stored in a geodatabase, we can set the geodatabase as the workspace. If the inputs are Shapefiles, TINs, etc., we can set the geodatabase as the folder where the input files are stored. On the contrary, if the workspace is not set, the input and output parameters need to be set using absolute paths. Even when workspace is set, we can still input or output any dataset outside the workspace by directly using the absolute path, for example:

   *arcpy.MakeFeatureLayer_management("I:\\sampledata\\data.gdb\\ bearMove", "inferLy")*

2. Open the output tables, and list the total lengths of roads in each polygon (Figure 9.7)

---

## 9.4 Accessing and Editing Data with Cursors

Data analysis or processes often require records to access field values. Use the cursor to point to a record in a table or feature class.

A cursor is a data access object that can be used either to iterate through the set of rows in a table or to insert new rows into a table. Cursors are commonly used to read and update attributes. Cursor is one of the ArcPy function in the Cursor data access module (requiring ArcGIS10.1+), and is recommended due to its high performance and easy operation.

Cursors have three forms. They are

- SearchCursor—Read-only access to obtain the geometry and attributes of feature records
- UpdateCursor—Read-write access to update feature records
- InsertCursor—Write access with capability to create new records

### 9.4.1 SearchCursor

The SearchCursor function establishes a read-only cursor on a feature class, such as a shapefile or a table. The SearchCursor can be used to iterate

```
"""
Set the path of the input data roads and source data. You need to change to your own path.
In this sample, the workspace is a geodatabase.
"bearMove" and "roads" are two feature classes in the geodatabase.
"""

arcpy.env.workspace = "O:\\Book\\Code\\9\\chp9Data\\bookSampleData.gdb"
# ensure bearMove is in workspace first
arcpy.MakeFeatureLayer_management("bearMove", "inferLy")
arcpy.MakeFeatureLayer_management("roads", "targetLy")

"""
"MakeFeatureLayer_management" can create a feature layer object from the path of the input,
which is a string. "SelectLayerByAttribute", "Clip_analysis", and "Statistics_analysis" are
then conducted on the feature layer.
"""

for i in range(0,9):
    # select the polygon with FID = i
    arcpy.SelectLayerByAttribute_management("inferLy", "NEW_SELECTION", "\"OBJECTID \"="+str(i))
    # execute clip analysis and out intermediate data "out_" + str(i) in workspace
    fc = arcpy.Clip_analysis("targetLy", "inferLy", "out_"+str(i))
    # execute sum statistical analysis and output result "sum_" + str(i) in workspace
    arcpy.Statistics_analysis(fc, "sum_" + str(i), [["Shape_Length", "SUM"]])
```

**CODE 9.2**

Automate calculating the road length within each polygon.

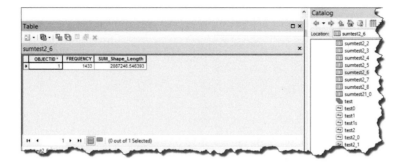

**FIGURE 9.7**
Example of output summary table generated by Code 9.2.

through row objects and extract field values. The syntax of SearchCursor is as follows:

*arcpy.da.searchCursor (in_table, field_names, {where_clause}, {spatial_reference}, {explode_to_points}, {sql_clause})*

The first argument (input feature table) and second argument (the queried fields) are required while others are optional (e.g., limited by a where clause or by field, and optionally sorted).

Hands-On Practice 9.5: Get Attributes of Each Feature of a Shapefile Using the SearchCursor

1. Open a Python window in ArcMap and type Code 9.3.

```
# change path according to your own
inputdata = "O:\\Book\\Code\\9\\chp9Data\\Partial_Streets.shp"

"""
    Open a SearchCursor and include a list of attribute(s) that you want to
    access (e.g. Shape_Leng, NAME, TYPE) in the parameter(s).
"""
rows = arcpy.da.SearchCursor(inputdata, ["Shape_Leng", "NAME", "TYPE"])
outputFile = open("C:\\ArcGISdata\\results.txt", "w")

# iterate through the rows in the cursor
for row in rows:
        # attributes are accessed using row[index] - e.g. row[0] is "Shape_Leng"
        outputFile.write("{}, {}, {}\n".format(row[0], row[1], row[2]))

"""
    The cursor will place a lock on the data until either the script completes or
    the cursor object is deleted.  Therefore, we need to delete the row and
    cursor objects to remove read locks on the data source.
"""
del row
del rows
outputFile.close()
```

**CODE 9.3**
SearchCursor with for statement.

```
inputdata = "O:\\Book\\Code\\9\\chp9Data\\Partial_Streets.shp"
with arcpy.da.SearchCursor(inputdata, ["Shape_Leng", "NAME", "TYPE"]) as rows:
        for row in rows:
                print "{}, {}, {}\n".format(row[0], row[1], row[2])
```

**CODE 9.4**
SearchCursor using with statement.

```
inputdata = "O:\\Book\\Code\\9\\chp9Data\\Partial_Streets.shp"

# select the features with FID < 10
with arcpy.da.SearchCursor(inputdata, ["Shape_Leng", "NAME", "TYPE"],
"FID < 10") as rows:
        for row in rows:
                print "{}, {}, {}\n".format(row[0], row[1], row[2])
```

**CODE 9.5**
SearchCursor with where clause ("FID < 10").

2. Check the results.txt file. What is included in the file? How many lines you can find in the file?

3. Search cursors also support the *with* statement. Using a *with* statement will guarantee that the iterator is closed and the database lock is released. By applying the *with* statement, the above code can be changed to Code 9.4.

4. A *where* clause may be used to limit the records returned by the cursor. Run Code 9.5 and check the result again. How many lines are included in the result file?

5. SearchCursor can also access the feature geometry. Run Code 9.6 and Code 9.7, and then check the result again:

```
with arcpy.da.SearchCursor(inputdata, ["SHAPE@", "SHAPE@LENGTH"],
"FID < 10") as rows:
        for row in rows:
            # return the x and y of the first point and the length
            of each feature
            print "{}, {}\n".format(row[0].firstPoint, row[1])
```

**CODE 9.6**
Accessing geometry using SearchCursor example 1.

```
with arcpy.da.SearchCursor(inputdata, ["SHAPE@"], "FID < 10") as rows:
        for row in rows:
            for pntarray in row[0]:
                for pnt in pntarray:
                    # return a tuple of x and y coordinates of the first two
                    # features in the data
                    print("{0}, {1}".format(pnt.X, pnt.Y))
```

**CODE 9.7**
Accessing geometry using SearchCursor example 2.

```
inputdata = "O:\\Book\\Code\\9\\chp9Data\\Partial_Streets.shp"

# create update cursor for the feature class
with arcpy.da.UpdateCursor(inputdata, ["Shape_Leng", "FID"]) as rows:
        for row in rows:
                # update the field "Shape_Leng" under the conditions
                of field "FID"
                if row[1] > 10:
                        row[0] = row[0] * 0.3048
                else:
                        row[0] = row[0] * 0.5
                rows.updateRow(row)
```

**CODE 9.8**
UpdateCursor example.

```
inputdata = "O:\\Book\\Code\\9\\chp9Data\\railway.shp"

# create update cursor for the feature class
with arcpy.da.UpdateCursor(inputdata, ["Shape", "FID"]) as rows:
        for row in rows:
                if row[1] < 10:
                        rows.deleteRow()
```

**CODE 9.9**
Using UpdateCursor to delete rows/records.

## 9.4.2 UpdateCursor

The UpdateCursor object can be used to update or delete specific rows in a feature class, shapefile, or table. The syntax of UpdateCursor is similar to that of SearchCursor:

*arcpy.da.UpdateCursor (in_table, field_names, {where_clause}, {spatial_reference}, {explode_to_points}, {sql_clause})*

Hands-On Practice 9.6: Update the Attributes of Each Feature for a Shapefile Using the UpdateCursor

1. Check the attribute value of "Shape_Leng" in the data "Partial_Streets.shp."
2. Open a Python window in ArcMap and type Code 9.8 in the editor.
3. Check the result "Partial_Streets.shp" to see whether the "Shape_Leng" attribute has been updated.
4. Open a Python window in ArcMap and type Code 9.9 in the editor and run the code to delete specific rows.

## 9.4.3 InsertCursor

InsertCursor is used to insert new records into a feature class, shapefile, or table. The InsertCursor returns an enumeration object, that is, a row in a table.

```
inputdata = "O:\\Book\\Code\\9\\chp9Data\\bookSampleData.gdb\\school"
# create the insert cursor and list the attributes that needs
to be filled up with values
cursor = arcpy.da.InsertCursor(inputdata, ["SCHOOL_NAM", "SHAPE@XY"])

"""
    Create the a new record with property "NAME" filled up
    with value "NewSchool" and xy coordinates filled up
    with (1847395.83394, 772277.97643)
"""
new_row = ["NewSchool", (1847395.83394, 772277.97643)]
cursor.insertRow(new_row)

# delete cursor to remove locks on the data
del cursor
```

**CODE 9.10**
Insert Cursor example.

Hands-On Practice 9.7: Inserts Rows into a Shapefile Using the InsertCursor

1. Open the attribute table of "school.shp" and count the number of records.
2. Open a Python window in ArcMap and type Code 9.10.
3. Check the new row added to the data "school.shp."

### 9.4.4 NumPy

As a new feature in ArcMap 10.1+, data access modules offer functions to enable the transformation between data array and feature classes or tables. Since NumPy library has powerful capabilities for handling arrays, the related function of ArcPy is developed based on NumPy. With arcpy. NumPyArrayToFeatureClass, arcpy.NumPyArrayToTable, and arcpy. TableToNumPyArray functions, users can quickly transform values that are organized in array into a feature class or table, and vice versa.

Without using the NumPy function in the data access module, include many more steps to create a point feature class so the performance is much lower (Code 9.11).

In contrast, creating the feature class with NumPy requires only one step. The arcpy.da.NumPyArrayToFeatureClass function actually creates a feature class and inserts two records inside.

Hands-On Practice 9.8: Create a Feature Class Using NumPyArrayToFeatureClass

1. Open the Python window in ArcMap, and run Code 9.12 to create a feature class using the NumPy functions in the data access module.

```
import arcpy
import numpy

# input array
array = numpy.array([[(1,(471316.3835861763, 5000488.782036674)),
        (2, (470402.49348005146, 5000049.216449278))],
        numpy.dtype([('idfield', numpy.int32),('XY','<f8',2)])))

# create the feature class with the field XY in the array
feat = arcpy.CreateFeatureclass_management
("O:\\Book\\Code\\9\\chp9Data\\Default.gdb", "out2", "POINT")
cursor = arcpy.da.InsertCursor(feat, ["SHAPE@XY"])

for i in array:
        new_row = [i[1]]
        cursor.insertRow(new_row)
del cursor
```

**CODE 9.11**
Add multiple new records to feature class using InsertCursor.

```
import arcpy
import numpy

output = "O:\\Book\\Code\\9\\chp9Data\\Default.gdb\\out"
# input array
array = numpy.array([[(1,(471316.3835861763, 5000488.782036674)),
            (2, (470402.49348005146, 5000049.216449278))],
            numpy.dtype([('idfield', numpy.int32),('XY','<f8',2)])))

# create the feature class with the field XY in the array
arcpy.da.NumPyArrayToFeatureClass(array, output, ['XY'])
```

**CODE 9.12**
Add multiple new records to feature class using NumPyArrayToFeatureClass.

2. Also run Code 9.11 in the Python window in ArcMap, and compare the execution time spent of with and without using numpy: the arcpy. da.NumPyArrayToFeatureClass has a much better performance.

## 9.5 Describing and Listing Objects

### 9.5.1 Describe

Describe a function used to read data properties by returning a property object. Depending on the arguments passed to the methods, the return object could be

- Data types (Shapefile, coverage, network datasets, etc.)
- Geometry type (point, polygon, line, etc.)

```
data = "O:\\Book\\Code\\9\\chp9Data\\bookSampleData.gdb\\railway"
dscb = arcpy.Describe(data)

if dscb.shapeType == "Polygon":
        print "I am polygon"
elif dscb.shapeType == "Polyline":
        print "I am polyline"
else:
        print "I am not either polyline or polygon"
```

**CODE 9.13**
Describe function example.

- Spatial reference
- Extent of features
- Path and so on

Hands-On Practice 9.9: Check the Properties of a Feature Class

1. Run Code 9.13 in ArcMap Python window and check the output. What is the shape type of the input feature class?
2. Replace the Describe parameter with another shapefile that has different geometric types and see how the results change when running it again.

## 9.5.2 List

ArcPy provides functions to list all data under a particular workspace or list corresponding information in data. ListFields are frequently used functions in ArcPy to list all the fields and associated properties of a feature class, shapefile, or table. Code 9.14 is an example of the ListFields function. The code controls operations to be conducted on specific fields only: those that are in "Double" type or that include the name "Flag."

Functions listing data under a workspace (e.g., ListDatasets, ListFeatureClasses, ListFiles, ListRasters, ListTables) are very useful to help batch processing. For example, to perform a buffer analysis for multiple

```
import arcpy

# list the field of the data "roads.shp" under the folder "ArcGISdata"
fieldlists = arcpy.ListFields("O:\\Book\\Code\\9\\chp9Data\\
bookSampleData.gdb\\roads")

# observe each field in the returning list
for field in fieldlists:
    # print out the property of the field, its name, scale, and type
    print field.name, field.scale, field.type
```

**CODE 9.14**
ListFields example.

polyline shapefiles in a workspace, list all the polyline shapefiles with the ListFeatureClasses method, then use a loop to go through each shapefile, performing a buffer analysis using Buffer_analysis method.

Hands-On Practice 9.10: Perform Buffer Analysis for Every Polyline Feature Class in a Workspace

1. Copy the data from Chapter 9 to your workspace (e.g., 'C:\\ sampledata'), or use any shapefiles you may have and put them under your workspace.
2. Run Code 9.15 and check the output console to answer *how many polyline features do you have under the workspace?*
3. Change the code and implement buffer analysis for every point feature class within 10 miles.

The Walk function in the data access module can help list all data under the workspace or under its sub-workspace in hierarchy. For example, there is a shapefile, a .png file, a geodatabase "Default.gdb," and a folder "temp" under a specific workspace "sampledata." This function will list all the files under "sampledata," the feature classes under "Default.gdb," and the files under "temp." The Walk function is much faster than traditional List functions. Code 9.16 is a sample code of the Walk function and Figure 9.8 shows its results.

```python
import arcpy

arcpy.env.workspace = "O:\\Book\\Code\\9\\chp9Data"
# get the list of all of the polyline feature classes
fcList = arcpy.ListFeatureClasses('*','Polyline')

# print the list of feature classes one at a time
for fc in fcList:
        print '-----Perform buffer analysis for Polyline:', fc
        inputFCName = fc[0:-4] # get rid of '.shp'
        outputFCName = inputFCName + '_buffer_10Meter' + '.shp'
        arcpy.Buffer_analysis(fc, outputFCName, '10 Meters')
```

**CODE 9.15**
List all polyline feature class and conduct buffer analysis on them.

```python
workspace = "O:\\Book\\Code\\9\\chp9Data"

for dirpath, dirnames, filenames in arcp.da.Walk(workspace):
        print "-------------"
        print dirpath;
        print dirnames;
        print filenames;
```

**CODE 9.16**
List files using arcpy.da.Walk function.

```
-------------
J:\course\sampledata
[u'Default.gdb', u'temp']
[u'greateryellowstoneecosystem4WD.png',
 u'pop_places2010_3states.shp']
-------------
J:\course\sampledata\Default.gdb
[]
[u'test']
-------------
J:\course\sampledata\temp
[]
[u'roadless_2004.shp', u'wild.shp']
```

**FIGURE 9.8**
Results of arcpy.da.Walk function.

## 9.6 Manipulating Complex Objects

Besides the commonly used simple objects, such as string (e.g., data path) or number (e.g., buffer distance), ArcPy offers classes to represent complex objects (e.g., spatial reference and attribute field).

One typical complex object is ValueTable. Many ArcTools include this object to allow input with multiple values. For example, the Union tool uses ValueTable as one input parameter (Figure 9.9). Accordingly, the script uses ValueTable to create the corresponding input parameter.

Another popular complex object, field, describes the field in the attribute table of a feature class, shapefile, and table (Code 9.17).

Meanwhile, ArcPy also provides the classes for manipulating geometry objects. Geometry information is usually stored in feature classes, feature layers, or datasets. Accessing geometry directly is not common when using

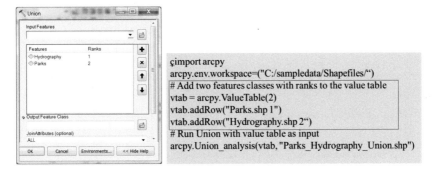

```
çimport arcpy
arcpy.env.workspace=("C:/sampledata/Shapefiles/")
# Add two features classes with ranks to the value table
vtab = arcpy.ValueTable(2)
vtab.addRow("Parks.shp 1")
vtab.addRow("Hydrography.shp 2")
# Run Union with value table as input
arcpy.Union_analysis(vtab, "Parks_Hydrography_Union.shp")
```

**FIGURE 9.9**
Perform union with ArcMap and python scripting.

```
import arcpy

fc = "O:\\Book\\Code\\9\\chp9Data\\bookSampleData.gdb\\roads"
desc = arcpy.Describe(fc)
# get a list of field objects from the describe object
fields = desc.fields

for field in fields:
        # manipulate field object and print out property of each field
        print field.name
        print field.aliasName
        print field.type
        if field.type == "Double":
                print field.scale
```

**CODE 9.17**
Accessing the properties of a field object.

geoprocessing tool for analysis. Sometimes, however, only locational information, such as creating sample points and making a buffer distance, is
needed. Using geometry objects can save the time on creating feature classes
that really do not persist. ArcPy offers several types of geometric objects:

- Point: a single point, which is the basic unit for all the other geometry
  types, cannot be directly used as input in geoprocessing functions
- Geometry: a general type
- Multipoint: a geometry entity with multiple points, consisting of
  Point(s)
- PointGeometry: a geometry entity with a single point, consisting of
  Point
- Polyline: a line geometry entity, consisting of Point(s)
- Polygon: a polygon geometry entity, consisting of Point(s)

Code 9.18 presents a buffer analysis with geometry as input.

Hands-On Practice 9.11: Manipulate Spatial Reference Object

```
# create a point
point = arcpy.Point(471316.38358618, 5000448)
# create the geometry interface of the point
pointgeom = arcpy.PointGeometry(point)
# create output geometry
outgeom = arcpy.Geometry()
# calculate the buffer of the create point geometry
arcpy.Buffer_analysis(pointgeom, outgeom, "5000 Meters")
```

**CODE 9.18**
Example of using geometry object to coduct buffer analysis.

```
# use the name of the coordinate system
spatialRef = arcpy.SpatialReference("Hawaii Albers Equal Area Conic")

# or use a projection file (.prj)
sr = arcpy.SpatialReference("C:\\coordsystems\\NAD 1983.prj")
```

**CODE 9.19**
Create SpatialReference object.

1. Create the object using a string as the path to a .prj file or using a string with the name of spatial reference (Code 9.19). SpatialReference is an ArcPy class.
2. Access the property of the spatial reference object (Code 9.20).
3. Create a Feature class in geodatabase using the spatial reference created (Code 9.21).
4. Create a Feature class in geodatabase using the spatial reference of another dataset (Code 9.22). The Describe function will be used to obtain the spatial reference information of the data.

```
print spatialRef.name
print spatialRef.XYTolerance
print spatialRef.metersPerUnit
print spatialRef.GCS
```

**CODE 9.20**
Access the properties of a SpatialReference object.

```
import arcpy

arcpy.env.workspace = ("O:\\Book\\Code\\9\\chp9Data")
# use the name of the coordinate system
spatialRef = arcpy.SpatialReference("Hawaii Albers Equal Area Conic")
# create the FDS using the spatialRef created from arcpy.
SpatialReference() method
arcpy.CreateFeatureDataset_management
('O:\\Book\\Code\\9\\chp9Data\\Default.gdb', 'results', spatialRef)
```

**CODE 9.21**
Create a feature class with a spatial reference.

```
import arcpy

arcpy.env.workspace = ("O:\\Book\\Code\\9\\chp9Data")
desc = arcpy.Describe('school.shp')
spatialRef = desc.SpatialReference

# create the FDS using the describe object's SR(SpatialReference) object
arcpy.CreateFeatureDataset_management('O:\\Book\\Code\\9\\chp9Data\\
Default.gdb','results', spatialRef)
```

**CODE 9.22**
Create a feature class with the spatial reference from another data.

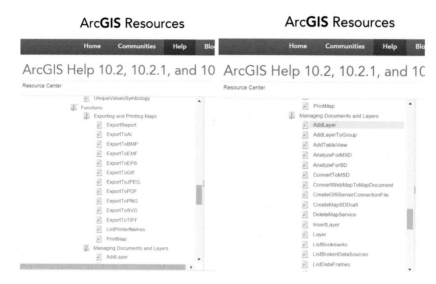

**FIGURE 9.10**
Functions in arcpy.mapping module.

## 9.7 Automating Map Production

Data manipulation and analysis are automated through Python scripts and ArcPy, while mapping processes involve user interactions on the interface or through scripting. ArcPy includes a mapping module, the major functions of which can be divided into two parts: exporting/printing maps and managing documents and layers. Figure 9.10 shows the functions in arcpy.mapping module.

Code 9.23 represents the process of making a map for a feature class using the predefined layer style. The process includes the opening of files and layers, changing layer style, and revising the location and content of map elements. The code is generalized from the script of an ArcTool named "CCMap Generator" (Teampython 2013). The script is open and can be downloaded from the Analysis and Geoprocessing community through the ArcGIS resource center website http://resources.arcgis.com/en/communities/analysis/.

## 9.8 Creating ArcTools from Scripts

The ArcMap geoprocessing framework allows users to create customized ArcTools from the Python scripts developed. By setting up scripts as ArcTools, users can configure input parameters and execute the functions through a graphic interface. This makes the user-defined function more

```
# open the map document, which is the *.mxd file
mxd = arcpy.mapping.MapDocument('CCMapDocTemplate.mxd')

# list the data frame in the map document - dfs is the first data frame in the document
dfs = arcpy.mapping.ListDataFrames(mxd)[0]

# create a layer from the dataset (e.g. a feature class) which will be styled and mapped
lyr = arcpy.mapping.Layer(featureclass)

# open the symbol style file *.lyr
symbollyrs = arcpy.mapping.Layer('CCMapSymbologyTemplate.lyr')

# get the first layer with name containing the string "test" inside the *.lyr file
symbollyr = arcpy.mapping.ListLayers(symbollyrs, ('*test*')) [0]

# change the symbol style of the feature class lyr in the dfs data frame into the pre-defined style symbollyr
arcpy.mapping.UpdateLayer(dfs, lyr, symbollyr, True)

# add the feature class with the updated symbol style into dfs data frame
arcpy.mapping.AddLayer(dfs, lyr)

# set the location and content of the first map element (e.g. text box) inside the mxd map document
elm = arcpy.mapping.ListLayoutElements(mxd, 'TEXT_ELEMENT', 'testelm') [0]
elm.elementPositionY = -1
elm.text = "this is a test map element"
elm.elementPositionX = 15

# export the map using resolution in 300 dpi
arcpy.mapping.ExportToPDF(mxd, out_map, resolution = 300)
```

CODE 9.23

Making maps with predefined symbol style automatically.

**FIGURE 9.11**
General steps to add scripts as tools.

user-friendly and sharable, as well as more accessible for individuals who do not know how to run a stand-alone script. Furthermore, the custom tool can also be incorporated into ModelBuilder to be integrated with other tools for implementing more advanced analysis processes.

ArcPy provides several getting and setting parameter functions for the custom Python scripts to accept user input. The most typical functions are as follows:

- GetParameter: Gets the value of the specified parameter from interface as an object (e.g., SpatialReference)
- GetParameterAsText: Gets the value of the specified parameter from interface as a String
- GeoParameterValue: Returns the default value of the specified parameter for the tool
- SetParameter: Sets a specified parameter property using an object
- SetParameterAsText: Sets a specified parameter property using a String

The steps to add scripts as a tool in the ArcToolBox are shown in Figure 9.11.

### Hands-On Practice 9.12: Build ArcTools from Scripts

1. Before creating the ArcTool, copy Code 9.24 and save as a *.py file.

```
import arcpy

"""
        The following is the buffer tool script, where the first
        argument is the input feature, the second argument is the
        output feature, and the third argument is the buffer
        distance
"""
inputFC = arcpy.GetParameterAsText(0)
outputFC = arcpy.GetParameterAsText(1)
bufferDist = arcpy.GetParameterAsText(3)

# perform buffer analysis
arcpy.Buffer_analysis(inputFC, outputFC, bufferDist)
```

**CODE 9.24**
A script with GetParameter functions to obtain input from ArcTool interface.

2. Open 'Catalog' in ArcMap, click 'ToolBoxes', right click "My Toolboxes," and select "New" to create a new tool box (Figure 9.12). Revise the "ToolBox.tbx" to a meaningful name, such as "MyGeoProcessor.tbx."

3. Add Toolset by right click, and select "New" -> Toolset (Figure 9.13). Change the name of "toolset," such as "Buffer."

4. Then, right click "Buffer," and select "Add" -> "Script" (Figure 9.14).

   Specify (1) basic info for the tool, such as name, label, descriptions, etc., (2) the script file path, and (3) input and output parameters, including types and names, etc. Click "Finish" to complete creating your customized ArcTool (Figure 9.15).

5. Run the newly created tool and see what happens.

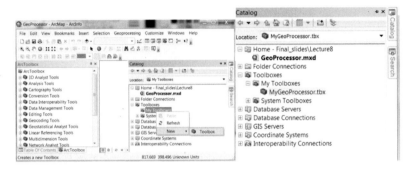

**FIGURE 9.12**
Add ArcToolBox.

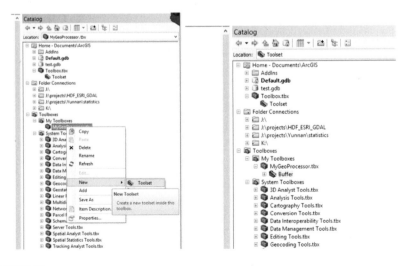

**FIGURE 9.13**
Add ArcToolset.

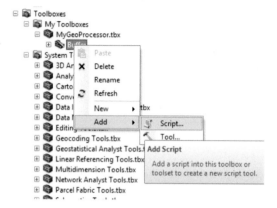

**FIGURE 9.14**
Add Python script as ArcTool.

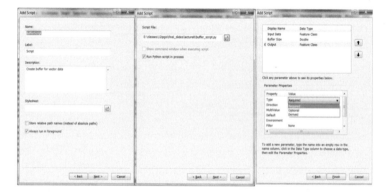

**FIGURE 9.15**
Configure the property of the tool, the path of script, and the input parameter.

## 9.9 Handling Errors and Messages

Informative messages help users track the running status of the script. ArcTools interface typically return three types of messages with the following symbols:

Informative messages during running of the script

Error message when a problem arises

Warning messages

- arcpy.GetMessage() method can return the geoprocessing messages. By default without any argument, it will get all types of messages. Using 0, 1, or 2 as argument, it will get the string of the information error, or warning ID messages separately:

**FIGURE 9.16**
Results of running script tool with messages added.

- GetMessages(): All messages
- GetMessages(0): Severity level 0 - Only informative messages
- GetMessages(1): Severity level 1 - Only warning messages
- GetMessages(2): Severity level 2 - Only error messages

The arcpy.AddMessage() method can be used to create geoprocessing messages (severity level = 0) for your scripts; the arcpy.AddWarning() method can be used to create warning messages (severity level =1) for your scripts; and the arcpy.AddError() method can be used to create error messages (severity level =2) for your scripts. As shown in Figure 9.16, the messages between "Start time: Tue …." and "Succeeded at …" are the ones added using the AddMessage() method.

Hands-On Practice 9.13: Add Message into the Custom Script Tool

1. Make a Python script with Code 9.25 and then add as an ArcTool.
2. Use the tool to test the buffer analysis and check the output.

## 9.10 External Document and Video Resources

The ArcGIS desktop help document and Esri online help document (Figure 9.1) should be the first place to check the syntax and sample codes of ArcPy classes, functions, and modules. By searching online, we can also find good code samples for many geoprocessing functions.

```
import arcpy

"""
        The following is the buffer tool script, where the first
        argumentis the input feature, the second argument is the
        output feature, and the third argument is the buffer distance
"""
inputFC = arcpy.GetParameterAsText(0)
arcpy.AddMessage('-------Input Feature: ' + inputFC)
outputFC = arcpy.GetParameterAsText(1)
arcpy.AddMessage('-------Output Feature: ' + outputFC)
bufferDist = arcpy.GetParameterAsText(2)
arcpy.AddMessage('-------Buffer Distance: ' + bufferDist)

# perform buffer analysis
arcpy.Buffer_analysis(inputFC, outputFC, bufferDist)
arcpy.AddMessage("Finished Successfully")
```

**CODE 9.25**
AddMessage examples.

The following resources could also provide the information from another perspective:

- Search for "Using Python in ArcGIS Desktop" in YouTube.
- Esri offers free Virtual Campus courses using Python in different version of ArcGIS Desktop, which can all be found by searching "Python" in the link below http://training.esri.com/gateway/index.cfm?fa=catalog.gateway&tab=0.
- ArcCafe: A blog, developed and maintained by the geoprocessing and analysis teams of Esri, which introduces Python scripts used to solve some common geoprocessing tasks.
- Esri usually provides sections to introduce the use of Python and ArcPy at their annual user conference. These videos can be found by searching "python" or "ArcPy" in the link http://video.esri.com/channel/2/events/series.

## 9.11 Implementing Spatial Relationship Calculations Using ArcGIS

Hands-On Practice 9.14: Calculate the centroid, perimeter, and area of polygons using arcpy.

Find the data "states.shp," and run Code 9.26 in the ArcMap Python window. Note that new fields must be added into the attribute table before the calculation in order to record the results (Figure 9.17).

```
# set the workspace
arcpy.env.workspace = "O:\\Book\\Code\\9\\chp9Data"

"""
    Add fields "centX", "centY", "polyArea", and "polyPeri" to record the calculated results.
    "DOUBLE" is the value type, 20 is the precision of the double type, and 10 is the scale.
"""

arcpy.AddField_management("states.shp","centX","DOUBLE",20,10)
arcpy.AddField_management("states.shp","centY","DOUBLE",20,10)
arcpy.AddField_management("states.shp","polyArea","DOUBLE",20,6)
arcpy.AddField_management("states.shp","polyPeri","DOUBLE",20,6)

"""
    Calculate the centroid, area, and perimeter using the CalculateField_management tool.
"PYTHON_9.3"
    means the calculation expression "!SHAPE.CENTROID.X!" is in Python 9.3 syntax.
"""

arcpy.CalculateField_management("states.shp","centX","!SHAPE.CENTROID.X!","PYTHON_9.3")
arcpy.CalculateField_management("states.shp","centY","!SHAPE.CENTROID.Y!","PYTHON_9.3")
arcpy.CalculateField_management("states.shp","polyArea","!SHAPE.AREA!","PYTHON_9.3")
arcpy.CalculateField_management("states.shp","polyPeri","!SHAPE.LENGTH!","PYTHON_9.3")
```

**CODE 9.26**
Calculate the centroid, perimeter, and area of polygons using arcpy.

| centX | centY | polyArea | polyPeri |
|---|---|---|---|
| -80.614059 | 38.641211 | 63086268312.096397 | 1793644.648034 |
| -78.851596 | 37.518636 | 103404844998.414 | 2999585.428349 |
| -77.015629 | 38.902603 | 171090249.453218 | 62054.401723 |
| -75.500554 | 38.995115 | 5317287218.88272 | 414744.898846 |
| -76.819383 | 39.064055 | 25226069378.6264 | 2119656.74972 |
| -77.802846 | 40.874889 | 117601507396.89101 | 1540291.740537 |
| -74.676628 | 40.198312 | 19429476786.153099 | 816773.016402 |

**FIGURE 9.17**
Result of Code 9.26.

Hands-On Practice 9.15: Selecting object based on spatial relationship (line intersection and point in polygon) using arcpy.

1. Find the two shapefiles "interstates" and "railway" in the disk, and select the interstate roads that intersect with railways. Selection will be conducted on the "interstate" features based on their spatial relationship with "railway" layer. Run Code 9.27 in ArcMap python window and Figure 9.18 shows the result.
2. Select the railway stations (in "amtk_sta.shp") in Virginia. Run Code 9.28 in ArcMap python window and see the result (Figure 9.19).

## 9.12 Summary

This chapter introduces programming within ArcGIS using Python scripts and ArcPy package. This chapter introduces

- Esri geoprocessing framework and ArcPy.
- The capability and syntax of ArcPy functions, classes, and modules.
- How to create simple or complex analysis workflows with ArcPy?
- How to manipulate vector data or objects through ArcPy?

```python
arcpy.env.workspace = "O:\\Book\\Code\\9\\chp9Data"

"""
    "MakeFeatureLayer_management" can create a feature layer object from the path of the input
    data, which is a string.  Selection will be conducted on the feature layer.
"""

arcpy.MakeFeatureLayer_management("interstates.shp", "roadLy")
arcpy.MakeFeatureLayer_management("railway.shp", "railLy")

# select the features in the interstates layer, which intersect with the features in the railway layer
arcpy.SelectLayerByLocation_management("roadLy", "INTERSECT", "railLy", selection_type="NEW_SELECTION")
```

CODE 9.27
Calculate line intersection.

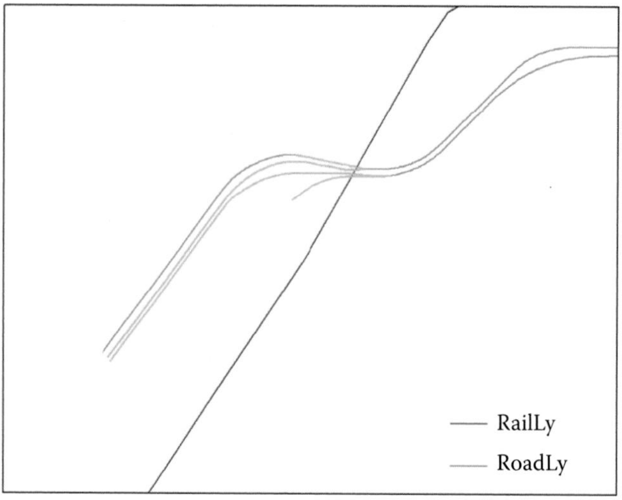

**FIGURE 9.18**
Line intersection result.

- How to create ArcTools from the python scripts?
- How to implement spatial calculations using ArcGIS?

## 9.13 Assignment

- Read the ArcPy section in ArcGIS desktop help or the online version.
- Find a road data or download data from the package of corresponding course material.
- Select highways from road data.
- Generate a buffer with 300 meters as the radius for the highway.
- Output the buffer as a transportation pollution zone.
- Add a field with the name of "buildings" with Long type in the buffer zone data.
- Count the number of buildings within each buffer zone and store into the new field.
- Write a report to explain how you conducted the analysis and programming.
- Compare the differences of implementing spatial calculations using ArcGIS and pure Python.

```
arcpy.env.workspace = "O:\\Book\\Code\\9\\chp9Data"

arcpy.MakeFeatureLayer_management("states.shp", "stateLy")
arcpy.MakeFeatureLayer_management("amtk_sta.shp", "stationLy")

# select Virginia from the state layer first
arcpy.SelectLayerByAttribute_management("stateLy","NEW_SELECTION",'"STATE_NAME"=\'Virginia\'')

# then select the railway stations (points) completely within Virginia (polygon)
arcpy.SelectLayerByLocation_management("stationLy","COMPLETELY_WITHIN","stateLy",selection_type="NEW_SELECTION")
```

CODE 9.28

Select all railway stations in Virginia.

**FIGURE 9.19**
Point in polygon result.

**NOTE:** All codes can be successfully executed on ArcGIS for desktop versions 10.2.2 through 10.3. There may be problems on running the code on more recent version of ArcGIS.

# 10

## Raster Data Algorithm

A raster is a data model representing geographic phenomena by pixels, and can be created using a variety of devices and techniques, such as digital cameras, scanners, coordinate-measuring machines, seismographic profiling, and airborne radar. This chapter introduces the raster concept, major categories of raster data, and how these data are displayed and stored in a computer. Basic raster data conversion and analysis methods are explored using three hands-on ArcGIS experiences.

### 10.1 Raster Data

With raster data model, geographic phenomena are represented as surfaces, regions, or segments. Therefore, this data model is based on the field view of the real world (Goodchild 1992). The field view is used widely for information organization in image analysis systems for resource- and environment-oriented applications. Raster data have two major categories: (1) discrete data, also called thematic or categorical data, as employed in land-use or soil maps; and (2) continuous data, also called nondiscrete data or surface data, as employed in Digital Elevation Models (DEMs), rainfall maps, or pollutant concentration maps.

A raster dataset represents geographic features in a 2D grid of cells known as picture elements (pixels) (Figure 10.1). The location of each cell is defined by its row and column numbers. The cell size dictates the spatial resolution of the data. The locations of geographic features are only represented by the nearest pixels. The value stored for each cell indicates the type of the object, phenomenon, or condition that is found in that particular location, and is normally given as the average value for the entire ground area covered by the pixel. Different types of values can be coded as integers, real numbers, or alphabet letters. If the code numbers are integers, then they are more likely referencing to nominal attributes (e.g., names in an associated table). Different attributes at the same cell location are each stored as separate themes or layers. For example, raster data pertaining to the soil type, forest cover, and slope covering the same area are stored in separate soil type, forest cover, and slope layers, respectively.

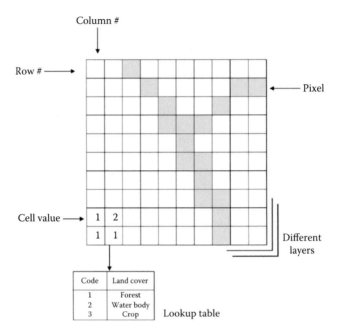

**FIGURE 10.1**
Raster data structure.

## 10.2 Raster Storage and Compression

Raster data are normally stored row by row from the top left, as illustrated in Figure 10.2. This is the simplest way of storing and searching for raster data. However, a raster, when stored in a raw state with no compression, can be

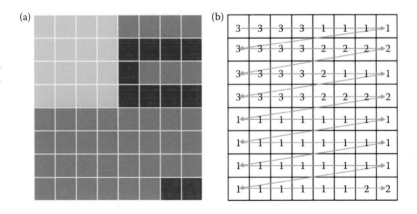

**FIGURE 10.2**
Raster image. (a) Each pixel of the raster is color coded and (b) value of each pixel and order of pixel storage.

extremely inefficient in terms of computer storage space. Taking Figure 10.2a, to extract the yellow area, the computer needs to extract the area one pixel at a time, following the storage order Figure 10.2b. After getting the four yellow cells in the first row, the computer needs to skip the remaining cells in the first row before reaching the targeted cells in the second row, which results in extra searching time.

Nowadays, data are increasingly available and has higher resolution. Computing capabilities have also improved. Therefore, better methods of data storage and compression are needed. Raster compression reduces the amount of disk space consumed by the data file, while retaining the maximum data quality. There are different methods for raster data compression, including Run Length Coding, quad tree coding, and others.

For multilayer raster datasets, the normal practice is to store the layers separately. It is also possible to store all information for each pixel together; however, this requires extra space to be allocated initially within each pixel's storage location for layers, which can be created later during analysis.

### 10.2.1 Run Length Coding

The Run Length Coding (Pountain 1987) is a widely used compression technique for raster data. The primary data elements are pairs of values or tuples, consisting of a pixel value and a repetition count, which specifies the number of pixels in the run. Data are built by reading each row in succession through the raster and creating a new tuple every time the pixel value changes or when the end of the row is reached. Figure 10.3 demonstrates the process of Run Length Coding. Suppose we have raster data stored in a 4 by

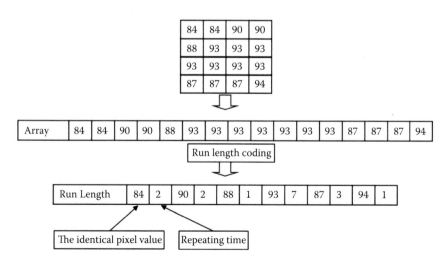

**FIGURE 10.3**
Example of Run Length Coding.

4 matrix. The computer will scan it starting from the top left and move right, working its way down, while keeping the data in an array. Then Run Length Coding will process the string of pixels into a string of pairs (the identical pixel value, and times of pixel repetition). The length of the initial string is 16 and after Run Length Coding the length is 12. Therefore, Run Length Coding effectively reduces the storage volume.

Run Length Coding has its limitations. For example, Run Length will not save storage in cases where pixel values do not repeat frequently. In some cases, such as DEM data in a mountainous area, neighboring pixels always have different values, and Run Length Coding may actually increase the length of the initial storage. However, Run Length Coding is very successful when dealing with black and white images, such as a fax. In this case, it is relatively efficient because most faxed documents are predominantly white space, with only occasional interruptions of black.

### 10.2.2 Quad Tree

The quad tree compression technique is the most common compression method applied to raster data (Gosselin and Georgiadis 2000). Quad tree coding stores the information by subdividing a square region into quadrants, each of which may be further divided into squares until the contents of the cells have the same values. Figure 10.4 demonstrates the process of quad tree compression. Suppose raster data is stored in a 4 by 4 matrix (Figure 10.4a). First, the quad tree divides the raster data into four square matrices (Figure 10.4b). In the sequence 0,1,2,3 (Figure 10.4b), the four matrices are checked on whether or not the contents of their cells have the same value. This process can be repeated recursively $n$ times, until the cells within a quadrant are all of the same value. For quadrant 0, the sequence is the same with previous levels of processes, and the division results in four other

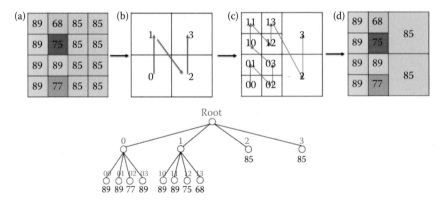

**FIGURE 10.4**

Quad tree process. (a) Pixel value of the rater, (b) search order of the four quadrants, (c) continuing dividing when finding non-equal values inside each quadrant of (b), and (d) final division.

squares, 00,01,02,03. The same process happens in quadrant 1. For quadrant 2 and 3, all cells have the same value, so no division is needed. Therefore, the final output of the quad tree is like Figure 10.4d. We call this arrangement a tree, whose nodes correspond to the squares. Nodes are connected if one of the corresponding squares immediately contains the other. The root node of the tree corresponds to the whole picture, the leaf nodes correspond to the single pixels.

Quad tree works for images with identical-value patterns. For each quad division used, four more storage elements are needed. Quad trees are of great interest for indexing spatial data, whereby cells that are adjacent in space are more likely to have similar spatial index addresses than in column or row ordering schema. Hence, data that are close together are also close in the storage system. This feature makes quad tree compressed data much easier and quicker to manipulate and access.

## 10.3  Raster Data Formats[*]

Raster format defines how the data are arranged and the corresponding compression type or level. Many data formats apply compression to the raster data so that all pixel values are not directly stored. Compression can reduce the data size to 30% or even 3% of its raw size, depending on the quality required and the method used. Compression can be lossless or lossy. With lossless compression, the original image can be recovered exactly. With lossy compression, the pixel representations cannot be recovered exactly. To implement compression, the raster file must include some head information about the compression method and parameters. Some raster formats, such as TIFF, GeoTIFF, IMG, and NetCDF, contain geoinformation, while others, such as BMP, SVG, and JPEG do not.

### 10.3.1  TIFF

TIFF (Tagged Image File Format) is an image format recognized by many computer systems. The TIFF imagery file format is used to store and transfer digital satellite imagery, scanned aerial photos, elevation models, scanned maps, or the results of many types of geographic analysis. TIFF supports various compression and tiling options to increase the efficiency of image transfer and utilization. The data inside TIFF files are categorized as lossless compressed or lossy compressed.

---

[*] "GIS file formats." Wikipedia: Raster. Wikipedia Foundation, Inc. http://en.wikipedia.org/wiki/GIS_file_formats#Raster.

## 10.3.2 GeoTIFF

GeoTIFF are TIFF files that have geographic (or cartographic) data embedded as tags within the TIFF file (Ritter and Ruth 1997). The geographic data can then be used to position the image in the correct location and geometry on the screen of a geographic information display. The potential additional information includes map projection, coordinate systems, ellipsoids, datums, and everything else necessary to establish the exact spatial reference for the file. Any Geographic Information System (GIS), Computer Aided Design (CAD), Image Processing, Desktop Mapping, or other type of systems using geographic images can read GeoTIFF files created on any system following the GeoTIFF specification.

## 10.3.3 IMG*

IMG files are produced using the IMAGINE image processing software created by ERDAS. IMG files can store both continuous and discrete, single-band and multiband data. These files use the ERDAS IMAGINE Hierarchical File Format (HFA) structure. An IMG file stores basic information including file information, ground control points, sensor information, and raster layers. Each raster layer in the image file contains information in addition to its data values. Information contained in the raster layer includes layer information, compression, attributes, and statistics. An IMG file can be compressed when imported into ERDAS IMAGINE, which normally uses the run length compression method (described in Section 10.2.1).

## 10.3.4 NetCDF

NetCDF (Network Common Data Form) is a set of software libraries and self-describing, machine-independent data formats that support the creation, access, and sharing of array-oriented scientific data (Rew and Davis 1990). It is commonly used in climatology, meteorology, and oceanography applications (e.g., weather forecasting, climate change) and GIS applications. It is an input/output format for many GIS applications, as well as for general scientific data exchange. NetCDF is stored in binary in open format with optional compression.

## 10.3.5 BMP

BMP (Windows Bitmap) supports graphic files inside the Microsoft Windows Operational System. Typically, BMP files data are not compressed, which can result in overly large files. The main advantages of this format are its simplicity and broad acceptance.

---

* "ERDAS IMAGINE .img Files." Purdue University. ftp://ftp.ecn.purdue.edu/jshan/86/help/html/appendices/erdas_imagine__img_files.htm.

### 10.3.6 SVG

Scalable Vector Graphics (SVG) are XML-based files formatted for 2D vector graphics. It utilizes a lossless data compression algorithm, and typically reduces data to 20%–50% of the original size.

### 10.3.7 JPEG

JPEG (Joint Photographic Experts Group) files store data in a format with loss compression (in major cases). Almost all digital cameras can save images in JPEG format, which supports eight bits per color for a total of 24 bits, usually producing small files. When the used compression is not high, the quality of the image is not as affected, however, JPEG files can suffer from noticeable degradations when edited and saved recurrently. For digital photos that need repeated editing or when small artifacts are unacceptable, lossless formats other than JPEG should be used. This format is also used as the compression algorithm for many PDF files that include images.

### 10.3.8 GIF

GIF (Graphic Interchange Format) is the first image format used on the World Wide Web. This format is limited to an 8-bit palette, or 256 colors. It utilizes lossless Lempel–Ziv–Welch (LZW) compression, which is based on patented compression technology.

### 10.3.9 PNG

PNG (Portable Network Graphic) is an open-source successor to GIF. In contrast to the 256 colors supported by GIF, this format supports true color (16 million colors). PNG outperforms other formats when large uniformly colored areas form an image. The lossless PNG format is more appropriate for the edition of figures and the lossy formats, as JPEG, are better for final distribution of photos, because JPEG files are smaller than PNG files.

## 10.4 Color Representation and Raster Rendering

### 10.4.1 Color Representation

Raster data can be displayed as either a grayscale or a color (RGB) image by transforming pixel values. A *colormap* is a lookup table used to translate each pixel's value into a color. A given pixel value is used as an index into the table, for example, a pixel value of nine will select the ninth element, or *colorcell* (Figure 10.5).

A grayscale image can be displayed on monochrome screens by transforming pixels into the intensity of gray level, which contains only a single value,

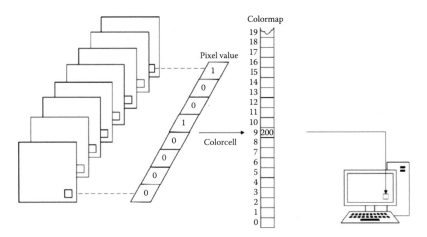

**FIGURE 10.5**
Pixel value to grayscale mapping.

in the colormap. Grayscale can be simulated on a color screen by making the red, green, and blue values equal in a given color cell, setting the brightness of gray pixels on the screen.

Most color screens are based on RGB color model. Each pixel on the screen is made up of three phosphors: one red, one green, and one blue, which are each sensitive to separate electron beams. When all three phosphors are fully illuminated, the pixel appears white to the human eye. When all three are dark, the pixel appears black. Each pixel value in the visible portions of a window is continuously read out of screen memory and looked up in the colormap. The RGB values in the specified colorcell control the intensity of the three primary colors and thus determine the color that is displayed at that point on the screen (Figure 10.6).

Raster data can also be transformed from RGB model to grayscale. The most common ways to conduct the transformation are Luster (Jack 2011), Intensity (Hoeffding 1963), and Luma (Bosch et al. 2007). The Luster method averages the most prominent and least prominent colors. The Intensity method simply averages the values. The Luma method is a more sophisticated version of the average method. It also averages the values, but it forms a weighted average to account for human perception; because human beings are more sensitive to green than other colors, green is weighted most heavily (Equation 10.1). Figure 10.7 shows the transformation results.

$$v' = \begin{cases} \dfrac{\max(R,G,B) + \min(R,G,B)}{2}, & \text{Luster method} \\[2ex] \dfrac{R+G+B}{3}, & \text{Intensity method} \\[2ex] 0.21 \times R + 0.72 \times G + 0.07 \times B, & \text{Luma method} \end{cases} \qquad (10.1)$$

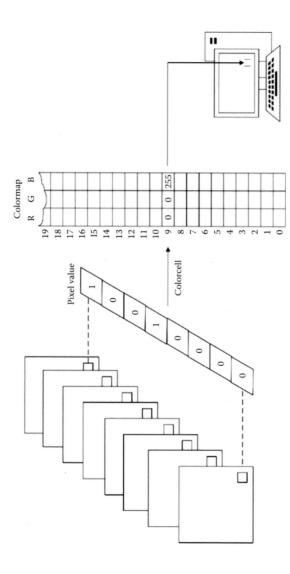

**FIGURE 10.6**
Pixel value to RGB mapping with the colormap.

Original image                 Luster method

Intensity method               Luma method

**FIGURE 10.7**
RGB to grayscale. (Original image from http://www.coloringpages456.com/color-pictures/.)

The *color depth* measures the amount of color information available to display or print each pixel of a digital image. Owing to the finite nature of storage capacity, a digital number is stored with a finite number of bits (binary digits). The number of bits determines the radiometric resolution of the image. A high color depth leads to more available colors, and consequently to a more accurate color representation. For example, a pixel with one bit depth has only two possible colors. A pixel with 8 bits depth has 256 possible color values, ranging from 0 to 255 (i.e., $2^8-1$), and a pixel with 24 bits depth has more than 16 million of possible color values, ranging from 0 to 16,777,215 (i.e., $2^{24}-1$). Usually, the color depths vary between 1 and 64 bits per pixel in digital images.

## 10.4.2 Raster Rendering

Raster datasets can be displayed, or rendered, on a map in many different ways. Rendering is the process of displaying the data. The map display depends on both the data itself and the chosen rendering method. Methods of rendering raster data commonly used in GIS software include Stretched, RGB Composite, Classified, Unique Values, and Discrete Color.

The *Stretched* renderer represents continuous data by stretching it on the statistics of the raster dataset. A stretch increases the visual contrast of the raster display, especially when the raster display appears dark or has little contrast. The image may not contain the entire range of values a computer can display (Figure 10.8a, b); therefore, by applying a contrast stretch, the image's

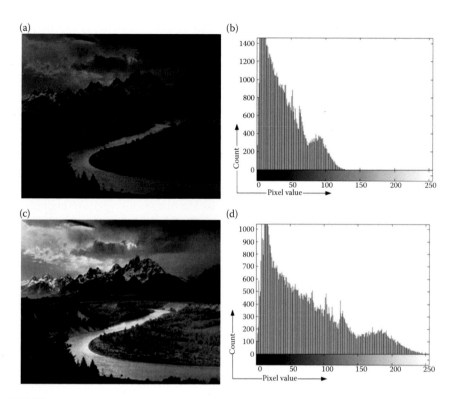

**FIGURE 10.8**
Stretch renderer. (a) Original figure, (b) Histogram of original figure, (c) Stretched figure, (d) Histogram of stretched figure.

values could be stretched to utilize this range (Figure 10.8d). In the case of eight bit planes, values are calculated in Equation 10.2.

$$\begin{cases} v' = m \times v + c \\ m = \dfrac{2^8 - 1}{\max(v) - \min(v)} \\ c = 2^8 - 1 - m \times \max(v) \end{cases} \tag{10.2}$$

where $v'$ refers to the stretched pixel value and $v$ refers to the original pixel value. This may result in a crisper image, and some features may become easier to distinguish (Figure 10.8c).

Different stretches will produce different results in the raster display; standard methods include Standard Deviation, Minimum–Maximum, Histogram Equalize, and Histogram Specification.

The *RGB Composite* renderer uses the same methods as the Stretched renderer, but allows combining bands as composites of red, green, and blue.

Multiband raster datasets, such as satellite or aerial imagery, are usually displayed in different combinations of bands using RGB Composite.

The *Classified* renderer is used with a single-band raster layer. The classified method displays thematic raster by grouping cell values into classes. Continuous phenomena such as slope, distance, and suitability can be classified into a small number of classes which can be assigned as colors. Classification is a spatial data operation used in conjunction with previous selection operations. This can either create a new dataset or present a different view of the same data (e.g., display properties).

Two less common renderers are *Unique Values* and *Discrete Color*. The *Unique Values* renderer is used to display each value in the raster layer individually. It is suitable for data containing discrete categories representing particular objects, such as a thematic map representing land use or types. The Unique Values renderer can display each value as a random color or use a color map. *Discrete Color* renderer displays the values in the raster dataset using a random color. This renderer is similar to the Unique Values renderer, but is more efficient when there are a large number of unique values because it does not have to calculate how many unique values exist. The Discrete Color renderer assigns a color to each unique value until it reaches the maximum number of colors chosen. The next unique value starts at the beginning of the color scheme; this process continues until each unique value has a color assigned to it.

## 10.5  Raster Analysis

Raster analysis includes various types of calculations based on pixels. In this section, we introduce several raster data analyses that are conducted frequently, including reclassification, overlay, and descriptive operations.

Reclassification is the process of reassigning new output values to a value, a range of values, or a list of values in a raster. It is conducted when (1) the value of a cell should be changed, for example, in the case of land change over time; (2) various types of values should be grouped together; or (3) specific values should be removed from the analysis. There are several approaches for reclassification, such as using lookup table and ranges of values.

Using a lookup table, the reclassification conducts a one-to-one change. For example, in Figure 10.9, to perform a habitat analysis, the pixel values on a land use raster, which represent numerous types of land use, need to be changed to represent a simple preference values of high, medium, and low (e.g., values 1, 2, and 3). The types of land most preferred are reclassified to higher values and those less preferred to lower values. For instance, forest is reclassified to 3, pasture land to 2, and low-density residential land to 1.

Using a *ranges of values* process, the reclassification is conducted in a many-to-one change, reclassifying a range of values to some alternative

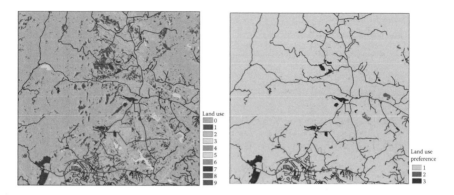

**FIGURE 10.9**
Reclassification of categorical data involves replacing individual values with new values. For example, land use values can be reclassified into preference values of low (1), medium (2), and high (3).

value and another range to a different alternative value. In our hypothetical analysis of habitat, the second layer in the suitability model is based on the preference for locations far from roads. Illustrated in Figure 10.10, a distance map (continuous data) is created from the existing roads theme. Instead of individually reclassifying each of the thousands of distance values on a 1-to-3 preference scale, the values can be divided into three groups. The farthest group receives the highest preference value, a value of 3, and the nearest group, a value of 1.

Descriptive operations can also be used to conduct raster analysis. Operations include minimum, maximum, mean, and median values, and can be operated on a different spatial scale such as local, focal, zonal, or global. Local operations work on individual raster cells, or pixels. Focal operations work on cells and their neighbors, whereas global operations work on the entire layer. Finally, zonal operations work on areas of cells that share the same value. In GIS software, operations can be conducted in raster calculator, where mathematical calculations and trigonometric functions are available.

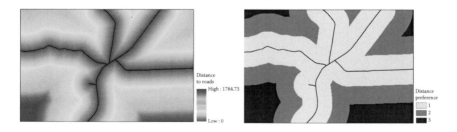

**FIGURE 10.10**
Reclassification of continuous data involves replacing a range of values with new values. For example, a raster depicting distance from roads can be reclassified into three distance zones.

Other applications of raster analysis can be DEM display, hydrology analysis (introduced in Chapter 12), zonal statistics, buffer analysis, and so on.

## 10.6 Hands-On Experience with ArcGIS

### 10.6.1 Hands-On Practice 10.1: Raster Color Renders

1. Figure 10.11a shows a land cover dataset stored in the geodatabase "chp10data.gdb." The raster is rendered using unique values with a prespecified color map. Observing the layer in the "Table of Contents" in ArcMap, when the label of this layer is shown as Figure 10.11b, the layer should contain a color map. Try to open the layer properties window (Figure 10.11c) by right clicking on the name of the layer in the "Table of Contents" and select the "Properties…," and select the "Symbology" tab in the window. The raster dataset rendered using color should have the choice of "Unique Values," "Discrete Color," and "Classified." The color map render is applied to the raster dataset with one band.

2. Change the color render mode. As shown in Figure 10.12a, export the land cover data. In the "Export Raster Data" window (Figure 10.12b), select "Use Renderer," which allows the output raster using the same color schema, and select "Force RGB," which will transfer the single band raster into the new Raster storing the pixel values in RGB mode.

When the raster is stored in RGB mode, we will see three sublayers in the "Table of Contents" (Figure 10.13) and under the "Symbology" tab in the "Layer Properties" window, there is an "RGB Composite" renderer choice, but "Unique Value," "Classified," and "Discrete Color" are no longer available.

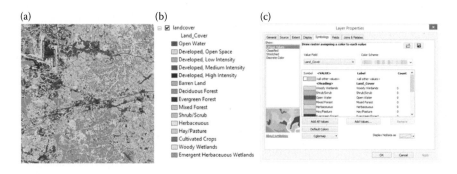

(a)　　　　　　　　　(b)　　　　　　　　　(c)

**FIGURE 10.11**

A raster data rendered using color map in ArcMap. (a) The raster layer rendered using color map. (b) The label of the layer. (c) The properties window.

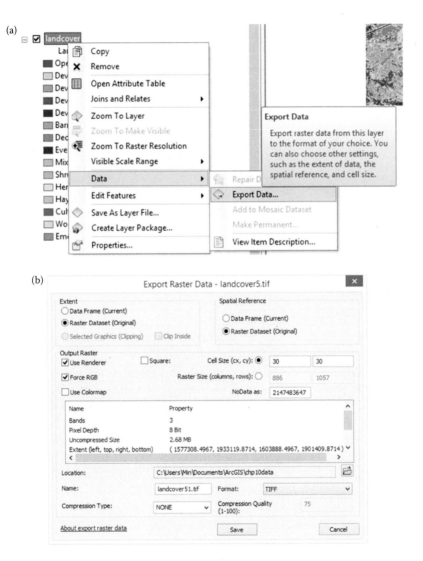

**FIGURE 10.12**
Steps of export raster data in ArcMap. (a) Export the land cover data and (b) "Export Raster Data" window.

### 10.6.2 Hands-On Practice 10.2: Raster Data Analysis: Find the Area with the Elevation Range between 60 and 100 and the Land Cover Type as "Forest"

1. In the chp10data.gbd, there are two raster datasets "dem" and "land-cover." Run Code 10.1 in ArcMap python window to classify the DEM data into several elevation ranges (0–10, 10–30, 30–60, 60–100, and >100 degrees). The result is shown in Figure 10.14.

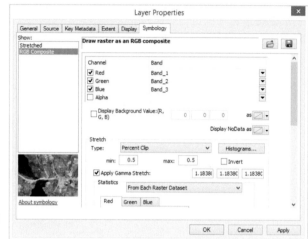

**FIGURE 10.13**
Raster displayed in RGB.

Note that the "Reclassify" function requires that the ArcMap users have the "Spatial Analyst" extension; therefore, check the license of the extension before executing the analysis.

The Reclassify function provides two methods for defining the classes: RemapRange redefines a range of values into a new class value; RemapValue defines the one-to-one mapping relationship, that is, replacing a single original value with a single new value.

2. Land cover dataset stores the land cover types in detail. For example, "Forest" is divided into three subtypes: "Deciduous Forest," "Evergreen Forest," and "Mixed Forest." Run Code 10.2 in the ArcMap python window to generalize the land cover dataset. The result is shown in Figure 10.15.

3. Run Code 10.3 in ArcMap python window to overlay the classified DEM and land cover datasets to find the area in forest (land cover dataset pixel value = 4) with an elevation between 60 and 100 (DEM dataset pixel value = 100). For this specific dataset, one way to find the expected area is to add the two layers and find the pixels with the value in 104, then reclassify all pixels with the value in 104 into one class (1), and all other pixels into another class (0). Figure 10.16 shows the result of reclassification.

### 10.6.3 Hands-On Practice 10.3. Access the Attribute Information of Raster Dataset and Calculate the Area

1. Run Code 10.4 in ArcMap Python window to calculate the area of the raster dataset. The raster dataset should be projected already.

```
# classify elevations into classes
# set workspace
arcpy.env.workspace = "C:\\ArcGISdata\\chp10data\\chp10data.gdb"

"""
        Check the license of the spatial analyst extension.  The
        returning value "available" means the functions in this
        extension are usable.
"""
arcpy.CheckExtension("spatial")

"""
        Define the input data.  "Dem" is a file geodatabase raster
        dataset stored in "chp10data.gdb"
"""
inRaster = "dem"

"""
        Define the value classes.  The first two elements in the
        bracket [0,10,10] means the minimum and maximum values in the
        class and the third element means the new value for the pixels
        following in the class.
"""
ranges = arcpy.sa.RemapRange([[0,10,10],[10,30,30],[30,60,60],
[60,100,100],[100,175,175]])

"""
        Execute the Reclassify function based on the "Value" field of
        the raster using the RemapRange and set missing values as
        "NODATA".
"""
outDEM = arcpy.sa.Reclassify(inRaster, "Value", ranges, "NODATA")

"""
        Output the reclassify function result as a new raster dataset
        in the file geodatabase, named as "classifiedElevation"
        (Figure x left).
"""
outDEM.save("classifiedElevation")
```

**CODE 10.1**
Classify DEM data.

2. Using the classified land cover dataset that resulted from the previous practice, run Code 10.5 in the ArcMap Python window to calculate the area of each land cover type. The area of each land cover type can be first calculated using the proportion of the pixel counts of this type to the total pixel count, and then multiply the total area of the raster dataset. Accordingly, use the SearchCursor (see Chapter 8) to capture the counts of pixels.

**FIGURE 10.14**
Result of reclassifying DEM data in Code 10.1.

```
# generalize land cover types
# input data
inLandcover = "landcover"

"""
        Set the mapping relationship between old values and new values.
        The first element in the bracket [11,1] means the old value and
        the second element means the new value.
"""
values = arcpy.sa.RemapValue([[11,1],[21,2],[22,2],[23,2],[24,2],[31,3],
            [41,4],[42,4],[43,4],[52,5],[71,7],[81,8],[82,8],[90,9],
            [95,9]])

# execute the reclassify function and save as a new dataset
"classifiedLandcover"
outLandcover = arcpy.sa.Reclassify(inLandcover,"Value",values)
outLandcover.save("classifiedLandcover")
```

**CODE 10.2**
Generalize the land cover dataset.

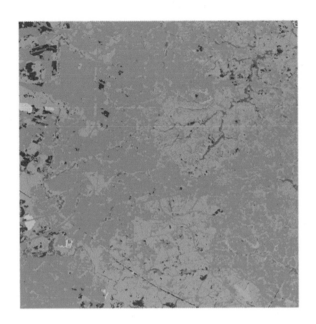

**FIGURE 10.15**
Result of reclassifying land cover type data in Code 10.2.

```
"""
        Use the raster calculator to add two layers.  Note that the
        raster calculator can execute many other algebra calculations
        on the
        raster dataset.
"""
temp = arcpy.gp.RasterCalculator_sa("'outLandcover' + 'outDEM'",
"overlayRaster")

"""
        Reclassify the layers into two classes - 1 and 0.  1
        represents the area that is in the forest and has expected
        elevation.
"""
outReclassify = arcpy.sa.Reclassify("overlayRaster", "Value", arcpy.
                                    sa.RemapRange([[12,103,0],
                                    [104,104,1],[105,184,0]]),
                                    "NODATA")
```

**CODE 10.3**
Reclassify and find specific areas.

**FIGURE 10.16**
Result of reclassifying land cover type data in Code 10.3.

```
# describe spatial extent of dataset and calculate area
desc = arcpy.Describe("classifiedLandcover")
area = desc.width * desc.height
```

**CODE 10.4**
Calculate area.

```
# describe spatial extent of dataset and calculate area
desc = arcpy.Describe("classifiedLandcover")
area = desc.width * desc.height

# store the total count of the pixels in the dataset
totalCount = 0
#initialize an array to put the counts of pixels in each land cover type
counts = []

"""
    Use the SearchCursor to access the Value and Count fields.
    The Value field is the land cover type value.
"""
with arcpy.da.SearchCursor("classifiedLandcover", ["Value", "Count"])
as \ cursor:
        for row in cursor:
                totalCount = totalCount + row[1]
                counts.append({'type': row[0], 'count':row[1]})

# calculate and print the area of each land cover type
for ele in counts:
        print 'The area of landcover type {0} is: {1}'.
        format(ele['type'],ele['count']/totalCount*area)
```

**CODE 10.5**
Calculate total area.

## 10.7 Chapter Summary

This chapter introduces raster data processing algorithms and demonstrates them in ArcGIS using Arcpy programming, which includes the following:

1. Raster data structure
2. Color representation
3. Raster data storage
4. Raster data compression
5. Raster data formats
6. Raster analysis
7. Hands-on experience with arcpy

### PROBLEMS

For raster data of your choice, design a scenario that requires reclassification. Explain the reasoning for reclassification and determine the purpose for the new classes. Calculate the area for each class and use different color rendering methods to present the result.

NOTE: All codes can be successfully executed on ArcGIS for desktop versions 10.2.2 to 10.3. There may be problem on running the code on more recent version of ArcGIS.

# 11

## Network Data Algorithms

A network is a system of interconnected elements, such as edges (lines) and connecting junctions (points) that represent possible routes from one location to another. People, resources, and goods tend to travel along networks: cars and trucks travel on roads, airliners fly on predetermined flight paths, oil flows in pipelines. By modeling potential travel paths with a network, it is possible to perform analyses related to the movement of the oil, trucks, or other agents on the network. The most common network analysis is finding the shortest path between two points (McCoy et al., 2001). This chapter introduces network representation, algorithms, and applications in GIS, and provides hands-on experience with ArcGIS.

## 11.1 Network Representation

### 11.1.1 Basics Network Representation

A network, sometimes called as graph, is a group or system of interconnected objects. It could be the transportation system, or a number of interconnected computers, machines, or operations. There are two essential components in a network: edges and nodes (Figure 11.1).

Each point in a network is called a vertex or node. The connections between vertices are referred to as edges or links. Mathematically, a network consists of a collection $V$ of vertices and a collection $E$ of edges. Each edge $e \in E$ is said to join two vertices, which are called its end points. When there is an edge $e$ that joins vertices $u$ and $v$, they are termed *adjacent*, and edge $e = (u,v)$ is said to be incident with vertices $u$ and $v$, respectively. For example, the network in Figure 11.2 can be represented by the following:

- $V = \{v_1, v_2, v_3, v_4\}$
- $E = \{(v_1, v_2), (v_2, v_4), (v_1, v_3), (v_1, v_4), (v_3, v_4)\}$

### 11.1.2 Directed and Undirected Networks

There are two types of network systems: directed networks and undirected networks. A directed network is a network in which the edges have directions,

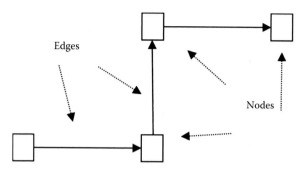

**FIGURE 11.1**
An example of basic network elements.

meaning that there is no distinction between the two vertices associated with each edge. For example, Figure 11.2 is actually an undirected network. Undirected networks are used to model networks in airlines, shipping lanes, and transit routes. For example, a two-way flight path connecting a set of cities can be represented as an undirected graph. The cities could be defined by the vertices and the unordered edges can represent two-way flight paths that connect the cities.

In contrast, edges in a directed network direct one vertex to another. In directed networks, an edge $e$ is defined by ordered pairs such as $<x,y>$ where vertex $x$ is the origin, and vertex $y$ is the destination. For example, edge $<V_2,V_3>$ is directed from $V_2$ to $V_3$. The directed network example in Figure 11.3a can be represented as $V = \{<V_1,V_2>, <V_3,V_2>, <V_3,V_1>\}$. Directed networks are used to model road, electricity, telephone, cable, sewer, and water systems. For example, a road network that connects a set of locations in a city with one-way roads can be represented as a directed graph. The locations can be represented by the vertices and the directed edges can represent the roads that connect the locations considering the traffic flow. Note whether to define a network as a directed one or not largely depends on the problem you are trying to model.

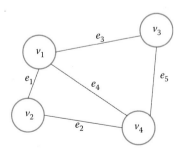

**FIGURE 11.2**
An example of basic network representation.

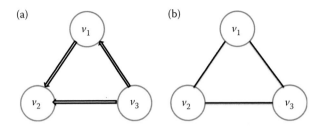

**FIGURE 11.3**
Example of directed and undirected network. (a) Directed network, (b) Undirected network.

### 11.1.3 The Adjacency Matrix

There are many ways to represent a network. One of the most popular ways is to use an adjacency matrix. An adjacency matrix is a square matrix used to represent a finite graph. The elements of the matrix indicate whether the pairs of vertices are adjacent or not in the graph. For example, for a network with $n$ vertices, the adjacency matrix for this network will be an $n \times n$ matrix, where $(i,j)$ is the index for the connection between vertex $i$ and $j$. If element $(i,j)$ is nonzero (i.e., "1"), it means that vertices $i$ and $j$ are connected, if it is zero it means they are not connected. Note that undirected adjacency matrices are symmetric. Figure 11.4 shows the difference between adjacency matrices for directed and undirected matrices.

### 11.1.4 Network Representation in GIS

Most GIS software uses node-edge representation for network analysis, which is a variant of adjacency matrix. Node-edge network representation

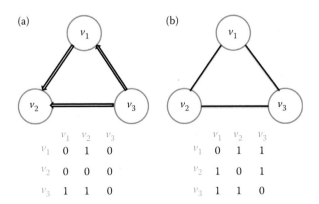

**FIGURE 11.4**
Adjacency matrix of directed and undirected network. (a) Directed network, (b) Undirected network.

**TABLE 11.1**

Example of Node Table for the Network in Figure 11.2

| ID | X | Y |
|----|-----|-----|
| $v_1$ | 23.2643 | 75.1245 |
| $v_2$ | 23.1443 | 74.1242 |
| $v_3$ | 23.2823 | 75.1315 |
| $v_4$ | 23.1442 | 75.1286 |

**TABLE 11.2**

Example of Link Table for the Network in Figure 11.2

| ID | Origin | Destination | One-Way |
|----|--------|-------------|---------|
| $e_1$ | $v_1$ | $v_2$ | Not |
| $e_2$ | $v_2$ | $v_4$ | Not |
| $e_3$ | $v_1$ | $v_3$ | Not |
| $e_4$ | $v_1$ | $v_4$ | Not |
| $e_5$ | $v_3$ | $v_4$ | Not |

maintains a table of nodes and a table of edges, as demonstrated in Tables 11.1 and 11.2.

- *Node table*: This table contains at least three fields: one to store a unique identifier and the others to store the node's $X$ and $Y$ coordinates. Although these coordinates can be defined by any Cartesian reference system, longitudes and latitudes ensure an easy portability to a GIS (Rodrigue, 2016).
- *Links table*: This table also contains at least three fields: one to store a unique identifier, one to store the node of origin, and one to store the node of destination. A fourth field can be used to state whether the link is unidirectional or not.

## 11.2 Finding the Shortest Path

### 11.2.1 Problem Statement

In graph theory, a path in a graph is a finite or infinite sequence of edges that connect a sequence of vertices, which, by most definitions, are all distinct from one another (Bondy and Murty, 1976). Finding the shortest path is one of the most important network analysis algorithms. It refers to finding the path between two vertices in a network that minimizes the cumulative distance (or weighted distance) of its constituent edges. For a regular shortest path problem, we usually make the following assumptions:

- Edges can be directed or undirected. A shortest path should respect the direction of its edges.
- The length of each edge does not have to be a spatial distance. It could also refer to time or some other cost.
- Distances between any nodes must be nonnegative.
- Not all vertices are reachable. If one vertex is not reachable from the other, there is no possible path between them.

### 11.2.2 A Brute Force Approach for the Shortest Path Algorithm

A brute force approach to finding the shortest path between two vertices in a network can be described as

- Step 1: Find all possible paths from the start point to the end point.
- Step 2: Calculate the length of each path.
- Step 3: Choose the shortest path by comparing the lengths of all different paths.

For example, given the network in Figure 11.5, we would like to find the shortest path from $A$ to all the other vertices. The number of each edge is the cost and Table 11.3 shows all the possible paths from $A$ to the other vertices. Although this method of finding the shortest path is simple and straightforward, the complexity of this approach increases exponentially with the number of vertices and edges. For example, if we connect $B$ and $C$, there will be at least two more routes from $A$ to $E$. In a real network application, we

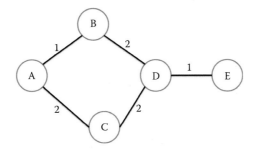

**FIGURE 11.5**
A network example used to illustrate finding shortest path problem.

**TABLE 11.3**

Brute Force Approach to Solving the Shortest Path Problem
(Find the Shortest Path from *A* to *E*)

| Destination Point | Possible Paths and Length | Shortest Path |
|---|---|---|
| B | AB: 1; ACDB: 6 | AB: 1 |
| C | AC: 2; ABDC: 5 | AC: 2 |
| D | ABD: 3; ACD: 4 | ABD: 2 |
| E | ABDE: 4; ACDE: 5 | ABDE: 4 |

usually have a large number of both vertices and edges (e.g., a transportation system), which would be very expensive and time-consuming from a computational standpoint. Therefore, a computationally efficient algorithm to calculate the shortest path is needed.

### 11.2.3 Dijkstra Algorithm

The Dijkstra algorithm is a widely used algorithm for solving the shortest path problem. It consists of the following six steps. Let the node at which we are starting be called the initial node. Let the distance of node $Y$ be the distance from the initial node to $Y$. Dijkstra's algorithm will assign some initial distance values and will try to improve them step by step (Dijkstra, 1959).

- Assign to every node a tentative distance value: set it to zero for the initial node and to infinity for all other nodes.
- Set the initial node as current. Mark all other nodes unvisited. Create a set of all the unvisited nodes called the unvisited set.
- For the current node, consider all of its unvisited neighbors and calculate their tentative distances. Compare the newly calculated tentative distance to the current assigned value and assign the smaller one. For example, if the current node $A$ is marked with a distance of 6, and the edge connecting it with a neighbor $B$ has length 2, then the distance to $B$ (through $A$) will be $6 + 2 = 8$. If $B$ was previously marked with a distance greater than 8, then change it to 8. Otherwise, keep the current value.
- After considering all of the neighbors of the current node, mark the current node as visited and remove it from the unvisited set. A visited node will never be checked again.
- If the destination node has been marked visited (when planning a route between two specific nodes) or if the smallest tentative distance among the nodes in the unvisited set is infinity (when planning a complete traversal; occurs when there is no connection

**TABLE 11.4**

Process of Using Dijkstra Algorithm to Solve the Shortest Path from $A$ to Other Vertices ($C_X$ Means the Cost from $A$ to $X$)

| Step | Selected Point | $C_A$ | $C_A$ | $C_A$ | $C_A$ | $C_A$ | Path | M |
|---|---|---|---|---|---|---|---|---|
| 1 | A | 0 | ∞ | ∞ | ∞ | ∞ | A | A |
| 2 | B | | 1 | 2 | ∞ | ∞ | AB | AB |
| 3 | C | | | 2 | 3 | ∞ | AC | ABC |
| 4 | D | | | | 3 | ∞ | ABD | ABCD |
| 5 | E | | 1 | 2 | 3 | 4 | ABDE | ABCDE |

between the initial node and remaining unvisited nodes), then stop. The algorithm has finished.

- Otherwise, select the unvisited node that is marked with the smallest tentative distance, set it as the new "current node," and go back to step 3.

As an example, the problem mentioned in the last section can be solved by using the Dijkstra algorithm (Table 11.4). The pseudo-code for the Dijkstra algorithm is shown in Table 11.5.

The implementation of Dijkstra algorithm is quite complex and requires different types of data structures and elaborate considerations. Therefore, ArcGIS examples are used to demonstrate how network analysis is supported in GIS. Source code is also available in many open-source software packages such as QGIS.

**TABLE 11.5**

Pseudo-Code for the Dijkstra Algorithm

```
Input: Network dataset (G), starting vertex (A)
{
DK1:   for each vertex v in G:
DK2:       dist[v] = ∞
DK3:   dist[A] := 0
DK4:   T = the set of all vertices in G
DK5:   while T is not empty:
DK6:       s = vertices in T with smallest dist[ ]
DK7:       delete s from T
DK8:       for each connected (neighbor) v of s:
DK9:           temp_Distance = dist[s] + dist_between(s, v)
DK10:          if temp_Distance < dist(v)
DK11:              dist [v] = temp_Distance
DK12:              shortest_Distance [v] = temp_Distance
DK13:  return shortest_Distance []
}
```

## 11.3 Types of Network Analysis

Network analysis allows you to solve common network problems, such as finding the best route across a city, finding the closest emergency vehicle or facility, identifying a service area around a location, servicing a set of orders with a fleet of vehicles, or choosing the best facilities to open or close. All of these problems could be solved by using Dijkstra algorithm or its variants.

### 11.3.1 Routing

Whether finding a simple route between two locations or one that visits several locations, people usually try to take the best route. But the "best route" can mean different things in different situations. The best route can be the quickest, shortest, or most scenic route, depending on the *impedance* chosen. The best route can be defined as the route that has the lowest impedance, where the impedance is chosen by the user. If the impedance is time, then the best route is the quickest route. Any valid network cost attribute can be used as the impedance when determining the best route.

### 11.3.2 Closest Facility

Finding the hospital nearest to an accident or the store closest to a customer's home address are examples of closest facility problems. When finding closest facilities, you can specify how many to find and whether the direction of travel is toward or away from them using GIS software. Once you have found the closest facilities, you can display the best route to or from them, return the travel cost for each route, and display directions to each facility. Additionally, you can specify an impedance cutoff for Network Analyst. For instance, you can set up a closest facility problem to search for restaurants within 15 minutes' drive time of the site of an accident. Any restaurants that take longer than 15 minutes to reach will not be included in the results (Figure 11.6).

### 11.3.3 Service Areas

With most network analysis tools, you can find service areas around any location on a network. A network service area is a region that encompasses all accessible streets, that is, streets that lie within a specified impedance. For instance, the 10-minute service area for a facility includes all the streets that can be reached within 10 minutes from that facility. One simple way to evaluate accessibility is by a buffer distance around a point. For example, find out how many customers live within a 5-kilometer radius of a site

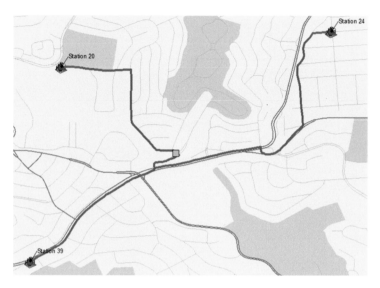

**FIGURE 11.6**
Example of closest facility.

using a simple circle. Considering that people travel by road, however, this method will not reflect the actual accessibility to the site. Service networks computed by Network Analyst can overcome this limitation by identifying the accessible streets within 5 kilometers of a site via the road network (Figure 11.7).

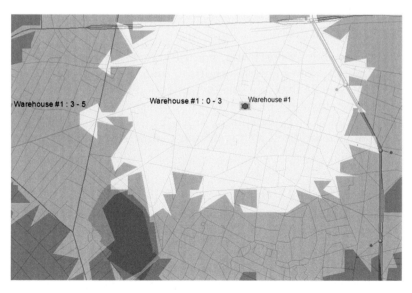

**FIGURE 11.7**
Example of service areas.

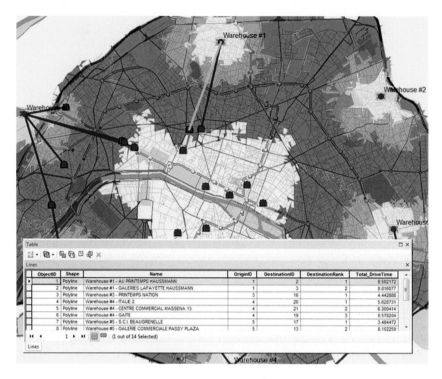

**FIGURE 11.8**
Example of OD cost matrix.

## 11.3.4 OD Cost Matrix

With most network analysis tool, you can create an origin–destination (OD) cost matrix from multiple origins to multiple destinations. An OD cost matrix is a table that contains the network impedance from each origin to each destination. Additionally, it ranks the destinations to which each origin connects in ascending order, based on the minimum network impedance required to travel from that origin to each destination (Figure 11.8).

## 11.3.5 Vehicle Routing Problem

A dispatcher managing a fleet of vehicles is often required to make decisions about vehicle routing. One such decision involves how to best assign a group of customers to a fleet of vehicles and to sequence and schedule their visits. The objective in solving such vehicle routing problems (VRP) is to provide timely customer service while keeping the overall operating and investment costs for each route to a minimum. The constraints are to complete the routes with available resources and within the time

**FIGURE 11.9**
Example of vehicle routing problem.

limits imposed by driver work shifts, driving speeds, and customer commitments (Figure 11.9).

### 11.3.6 Location-Allocation

Location-allocation could help choose, given a set of facilities, from which specific facilities to operate, based on their potential interaction with demand points. The objective may be to minimize the overall distance between demand points and facilities, maximize the number of demand points covered within a certain distance of facilities, maximize an apportioned amount of demand that decays with increasing distance from a facility, or maximize the amount of demand captured in an environment of friendly and competing facilities.

Figure 11.10 shows the results of a location-allocation analysis meant to determine which fire stations are redundant. The following information was provided to the solver: an array of fire stations (facilities), street midpoints (demand points), and a maximum allowable response time. The response time is the time it takes firefighters to reach a given location. In this example, the location-allocation solver determined that the fire department could close several fire stations and still maintain a 3-minute response time.

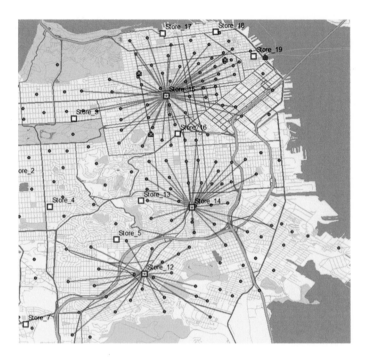

**FIGURE 11.10**
Example of location-allocation.

## 11.4 Hands-On Experience with ArcGIS

*Hands-On Practice 11.1: Run the codes in the ArcMap Python window step by step to help a salesman find the shortest travel path for calling on a number of customers.*

1. The network dataset and the stops of the salesman are available in the data disk under the folder "chp11data." The first step to conduct network analysis is to find the shortest path by using the arcpy. na.MakeRouteLayer function to create a route layer from the network dataset (Code 11.1).

   The input parameter is the ND layer of the network dataset. Give a name to your output route layer, such as myRoute. In this example, the impedance attribute is "Length." "FIND_BEST_ORDER" and "PRESERVE_BOTH" mean that the order of the salesman's stops can be changed when analyzing the shortest path (to approach an optimal result), but the first and end stops are preserved as his fixed start and end locations. The total length of the resulting path is calculated for reference (accumulate_attribute_name = "Length"). Figure 11.11 shows the route data layer in ArcMap created by Code 11.1.

```
#create a route layer from the network dataset

arcpy.env.workspace = 'C:\\ArcGISdata\\chp11data'

routeLy = arcpy.na.MakeRouteLayer(in_network_dataset = "roads_ND.nd",
out_network_analysis_layer =
"myRoute", impedance_attribute = "Length", find_best_order = "FIND_
BEST_ORDER",
ordering_type = "PRESERVE_BOTH", time_windows = "NO_TIMEWINDOWS",
accumulate_attribute_name = "Length", UTurn_policy = "ALLOW_UTURNS",
restriction_attribute_name = "#",
hierarchy = "NO_HIERARCHY", hierarchy_settings = "#",
output_path_shape = "TRUE_LINES_WITH_MEASURES", start_date_time =
"#").getOutput(0)
```

**CODE 11.1**
Script to create a route layer from the network dataset.

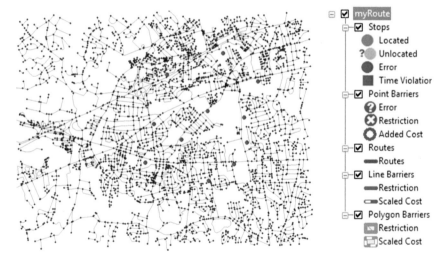

**FIGURE 11.11**
Network dataset (left) and the route layer generated using the dataset (right).

```
# get the sub classes of the route layer
naClasses = arcpy.na.GetNAClassNames(routeLy, "INPUT")
```

**CODE 11.2**
Script to get all input sublayer in the route layer.

2. Add the stops of the salesman in the "stops.shp" to the route layer. Code 11.2 is provided to obtain all the subclasses in the route layer structure. The subclasses of a route layer include Barriers, PolygonBarriers, PolylineBarriers, Stops, and Routes. Except the "Routes" class, all the other classes are the input classes that allow users to input stops and barriers to restrict the network analysis.

```
>>> naClasses = arcpy.na.GetNAClassNames(routeLy,
"INPUT")
>>> naClasses
{u'Barriers': u'Point Barriers',
 u'PolygonBarriers': u'Polygon Barriers',
 u'PolylineBarriers': u'Line Barriers', u'Stops':
 u'Stops'}
```

**FIGURE 11.12**
Returned result of Code 11.2: all input sublayer in the route layer.

The analysis output will be automatically stored in the "Routes" class. Since we set a parameter "INPUT" in the code below, the code will only return those input subclasses (Figure 11.12).

Run Code 11.3 to input stops in the "Stops" subclass. The default value of speed and length at the stops is set as 0, which means the salesman will not have speed and have length cost at those stops. The results of adding stops to each route layer are shown in Figure 11.13.

3. Run the Solve function (Code 11.4) to calculate the shortest path with "Length" as impedance (i.e., constraining cost).

Figure 11.14 (upper) shows the resulting route. The order of the stops is different from the ones in Figure 11.11, because the order has been changed to optimize the analysis result. The total length of the route has been accumulated and stored in the attribute table of the resulting route (Figure 11.14 lower).

```
"""
        Add stops (the points in the "stops.shp") to the "Stops"
        class in the route layer.
        Fieldmapping is used to input the attribute in the stops.shp
        to the "Stops"  subclass to constrain the network analysis
"""

fieldMappings = arcpy.na.NAClassFieldMappings(routeLy,
naClasses["Stops"])

# set the default value for the properties of the fieldmapping

fieldMappings["Attr_Length"].defaultValue = 0

fieldMappings["Attr_speed"].defaultValue = 0

# add the points in stops feature class into the sublayer "Stops"
of route layer with field mapping

arcpy.na.AddLocations(routeLy, "Stops", 'stops.shp', fieldMappings)
```

**CODE 11.3**
Script to add the salesman's stops to the "Stops" sublayer in the route layer.

**FIGURE 11.13**
Returned result of Code 11.3: the route layer with stops added in.

```
# create route for those stops
arcpy.na.Solve(routeLy)
```

**CODE 11.4**
Script to execute shortest path algorithm in the route layer.

4. In the above example, use "Length" as the impedance to calculate the shortest path. Now change the impedance parameter setting in Code 11.1 into "impedance_attribute = 'speed'" and run Codes 11.1 through 11.4 again to analyze the shortest path with road speeds as the impedance cost.

## 11.5 Chapter Summary

This chapter introduces network-related data representations and algorithms. It also demonstrates how to program the functions provided by ArcGIS through arcpy scripting:

1. Basic network representation
2. Directed and undirected networks

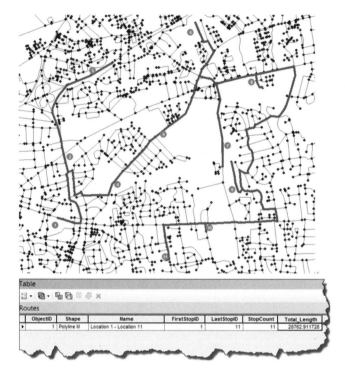

**FIGURE 11.14**
Routing result (upper: the route line; lower: the attribute table of the route layer).

3. Adjacency matrix
4. Links Table and Node Table
5. Shortest path algorithms (the Dijkstra algorithm)
6. Hands-on experience with ArcGIS through arcpy scripting

**PROBLEMS**

Review the class material and practice code, and develop a network for your University Campus for routing:

1. Capture the network data model.
2. Capture the datasets including vertices and links.
3. Implement a simple routing algorithm that can route from one point to another point.
4. Point can be buildings, parking lots, and other points of interest.

NOTE: All codes can be successfully executed on ArcGIS for desktop versions 10.2.2 through 10.3. There may be problem on running the code on more recent version of ArcGIS.

# 12

## Surface Data Algorithms

We live in a 3D space, where we observe entities stereoscopically. New advanced technologies have been developed to enable researchers to explore real-world 3D geospatial information. 3D is not only one of the key areas for GIS evolution but also the basis that guarantees the success of the popular Google Earth, Virtual Earth, and Image City. One of the major forms of 3D data is surface data. Surface data can represent surface features, such as elevation data, contamination concentrations, and water-table levels. This chapter introduces the fundamentals of surface data and related processing algorithms.

## 12.1 3D Surface and Data Model

### 12.1.1 Surface Data

Surface data is the digital representation of real or hypothetical features in 3D space. A surface is a representation of geographic features that can be considered as a $Z$ value for each location defined by $X$, $Y$ coordinates. Since a surface contains an infinite number of points, it is impossible to measure and record the $Z$ value at every point. Raster, triangulated irregular network (TIN), terrain, and Light Detection and Ranging (LiDAR) datasets are all surface data types. A surface is usually derived, or calculated, using specially designed algorithms that sample points, lines, or polygon data, and convert them into a digital surface.

### 12.1.2 Surface Data Model

3D surface data can be represented as discrete data or continuous data. Discrete data can be referred to as point-based or line-based, while continuous data represents field, nondiscrete surface data. Discrete data can be interpolated into continuous data.

#### 12.1.2.1 Discrete Data

Discrete data can be mass points, breaklines, and contour lines. Mass points are point data features that represent locations on the ground in three dimensions (ESRI 2016a,b). A mass point is defined by $x$, $y$, and $z$ coordinates, typically represented as an easting, a northing, and an elevation. Mass points are commonly used as components of digital elevation models and digital terrain models to represent terrain. Mass points are generally created in evenly spaced grid patterns, but may also be placed randomly depending on the method of creation and the characteristics of the terrain being defined.

Breaklines are line features that represent sudden changes in terrain, usually associated with linear ground features such as retaining walls, road edges, steep ridges, and ridgelines (ESRI 2016a,b). Like mass points, breaklines consist of vertices that are defined in $x$, $y$, and $z$ coordinates, typically represented as eastings, northings, and elevations. Breaklines are also commonly used as components of digital elevation models and digital terrain models to represent terrain.

Breaklines are classified into two groups: soft breaklines and hard breaklines. The Z values of soft breaklines are calculated from existing points, while those of hard breaklines are without calculation. In addition, soft breaklines are used to define boundaries that are not physical features of the landscape (e.g., TIN edges, political boundaries, vegetation, and soil types), so that each triangle will be assigned to one feature type. Hard breaklines, in contrast, are used to define interruptions in surface smoothness (e.g., streams, shorelines, dams, ridges, and building footprints).

Contours are commonly used to express digital elevation data; however, they can also be used to connect points of equal value for any such "surface" parameters, such as temperature, water table, or pollution concentrations. Each contour line corresponds to a specific value; therefore, contour lines never cross each other (Figure 12.1). Where there is less drastic change in

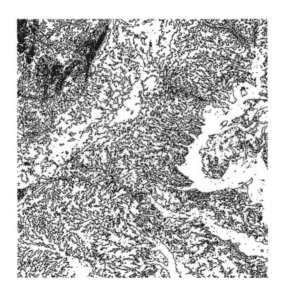

**FIGURE 12.1**
Contour lines.

values, the lines are spaced farther apart; where the values rise or fall rapidly, the lines are closer together. Contour lines can, therefore, be used not only to identify locations that have the same value, but also gradient of values. For topographic maps, contours are a useful surface representation, because they can simultaneously depict flat and steep areas (distance between contours) and ridges and valleys (converging and diverging polylines).

The elements needed to create a contour map include a *base contour* and a *contour interval* from values for a specific feature. For example, we can create a contour every 15 meters, starting at 10 meters. In this case, 10 meters would be considered the base contour and the contour interval would be 15 meters; the values to be contoured would be 10 m, 25 m, 40 m, 55 m, etc.

### 12.1.2.2 Continuous Data

Continuous surface data can be either Grid (e.g., Digital elevation model) or TIN. Grid and TIN are the most frequently used models in continuous surface representation, each offering its own advantages and shortcomings.

#### 12.1.2.2.1 Grid

Grid surface refers to a surface map plotted as a grid of surface values with uniformly spaced cells. This grid is in the same data structure as raster data, consisting of a rectangular matrix of cells represented in rows and columns. Each cell represents a defined square area on the Earth's surface and holds a value that is static across the entire cell (Figure 12.2). Elevation models are one such example of Grid surface models.

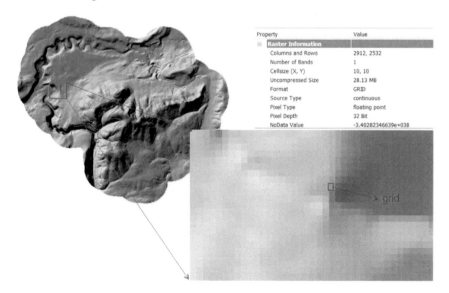

| Property | Value |
|---|---|
| **Raster Information** | |
| Columns and Rows | 2912, 2532 |
| Number of Bands | 1 |
| Cellsize (X, Y) | 10, 10 |
| Uncompressed Size | 28.13 MB |
| Format | GRID |
| Source Type | continuous |
| Pixel Type | floating point |
| Pixel Depth | 32 Bit |
| NoData Value | -3.40282346639e+038 |

**FIGURE 12.2**
Grid surface model.

The advantage of grid surfaces is their simplicity in terms of storage and processing. As a result, the computation of grid surface data is fast and straightforward; however, because the cell value is determined by interpolation, a grid surface has a fixed resolution as determined by the cell size. In addition, some source data may not be captured, which may cause loss of information.

### 12.1.2.2.2 TIN

TINs are a form of vector-based digital geographic data and are constructed by triangulating a set of vertices, each with its own $x$, $y$ coordinate and $z$ value (Figure 12.3). The vertices are connected with a series of edges to form a network of triangles. Triangles are nonoverlapping; thus, no vertex lies within the interior of any of the circumcircles of the triangles in the network (Figure 12.4). Each triangle comprises of three vertices in a sequence, and is adjacent to at least one neighboring triangle.

### 12.1.2.2.3 Comparison of Grid and TIN

TIN and Grid each has their advantages, so it would be arbitrary to credit one as being better than the other. Some advantages and disadvantages of TINs are as follows:

- The TIN model has variable resolution. A TIN preserves the $x$, $y$ location of input points, allowing for more detail where there are extensive surface variations and less detail where the changes are small or nonexistent.
- Since TINs are vector data, they can be displayed well at all zoom levels. Raster display degrades when you zoom in too close.

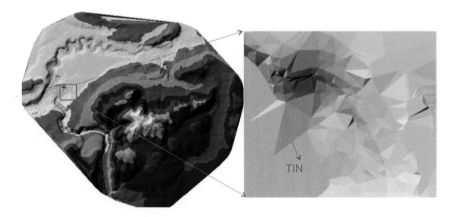

FIGURE 12.3
TIN.

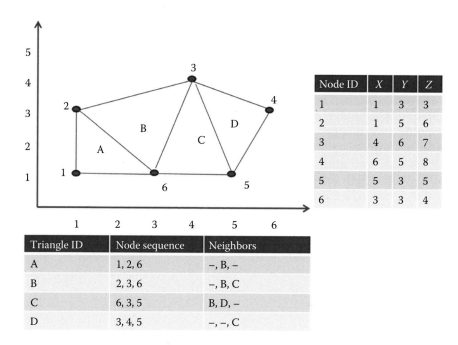

**FIGURE 12.4**
TIN triangles.

- For large-scale applications (those covering a small area in detail) or applications where display quality is very important, TINs are often a better choice.
- However, in many cases, TINs require visual inspection and manual control of the network.

On the other hand, Grids have their advantages and disadvantages:

- Grid data structure is straightforward and easy to process. Their matrix structure makes them well suited to analysis. A greater variety of mathematical and statistical functions are available for Grid versus TINs.
- Grid is a more familiar and readily available data type.
- For small-scale applications (those covering a large area), or applications that require statistical analysis of data, Grids are often a better choice.
- The disadvantage is the inability to use various grid sizes to reflect areas of different complexity of relief.

## 12.2 Create Surface Model Data

### 12.2.1 Create Grid Surface Model

Grid model data can be created from discrete features like mass points or contours by sampling a series of regularly spaced grid points with corresponding information and grid size. Then, based on points with known elevations (e.g., mass points and points on contour lines), a representative elevation for each grid can be interpolated or calculated. Each nearest point will have a weight based on the distance from that point to the grid. The used weight can be equal weight for all points or unequal weights for different points. Equal weight will calculate the elevation of the unknown point as the average of all points (Equation 12.1).

$$Z_0 = \frac{\sum\limits_{i=1}^{s} z_i}{s} \tag{12.1}$$

Unequal weights, such as exponential weight (weight $= e^{-d}$, where $d$ is the distance between Point $i$ and Point 0) or power weight (weight $= d^{-k}$, where $k$ is the power number), can provide more variability for different purposes. $Z_0$ is the estimated value at Point 0, and $Z_i$ is the $Z$ value at known Point $i$ (Equation 12.2).

$$Z_0 = \frac{\sum\limits_{i=1}^{s} \text{weight}_i * z_i}{\sum\limits_{i=1}^{s} \text{weight}_i} \tag{12.2}$$

Different methods exist for calculating the grid's elevation based on the weights' scheme and the number of nearest points used. Commonly used interpolation methods include inverse distance weighting (IDW), spline, kriging, and natural neighbors. IDW weights the points closer to the target cell more heavily than those farther away. Spline fits a minimum curvature surface through the input points. Kriging is a geostatistical interpolation technique in which the surrounding measured values are weighted to derive a predicted value for an unmeasured location. These weights are based on the distance between the measured points, the prediction locations, and the overall spatial arrangement among the measured points. Natural neighbors create a Delaunay triangulation of the input points, selecting the closest nodes that form a convex hull around the interpolation point, and then weighting their values proportional to their area. In ArcGIS, Grid surface is phrased as Raster.

## 12.2.2 Creating TIN Surface Model

A TIN surface can be created from discrete data, such as points, lines, and polygons that contain elevation information. Normally, mass points are the primary input to a TIN and determine the overall shape of the surface. Breaklines are used to enforce natural features, such as lakes, streams, ridges, and valleys. Polygon features are integrated into the triangulation as closed sequences of three or more triangle edges. It usually takes multiple steps to create TIN from points:

1. *Pick sample points.* In many cases, sample points must be selected from control points, such existing, dense Digital Elevation Model (DEM) or digitized contours, to ensure accuracy of representation. There are several existing algorithms for selecting from a DEM: the Fowler and Little (1979) algorithm, the VIP (Very Important Points) algorithm (Chen and Guevara 1987), and the Drop heuristic algorithm (Lee 1991). In essence, the intent of these methods is to select points at significant breaks of the surface.

2. *Connect points into triangles.* The selected TIN points will then become the vertices of the triangle network. Triangles with angles close to 60 degrees are preferred since this ensures that any point on the surface is as close as possible to a vertex. There are different methods of interpolation to form the triangles, such as Delaunay triangulation or distance ordering. Delaunay triangulation, the method most commonly used in practice, ensures that no vertex lies within the interior of any of the circumcircles of the triangles in the network (Figure 12.5). Delaunay triangulation is accomplished either by starting from the edges of the convex hull and working inward until the network is complete, or by connecting the closest pair that must be a Delaunay edge, searching for a third point such that no other point falls in the circle through them, and then working outward from these edges for the next closest point.

3. *Model the surface within each triangle.* Normally, the surface within each triangle is modeled as a plane.

## 12.2.3 Conversion between TIN and Raster Surface Models

The conversion from a TIN to a raster requires the determination of a cell size, which represents the horizontal accuracy. Based on the cell size, elevation values can then be interpolated from the TIN at regularly spaced intervals across the surface. With decreasing cell size, more points will be interpolated, yielding an output raster that resembles the input TIN more closely. A TIN's slope and aspect values can also be converted to raster.

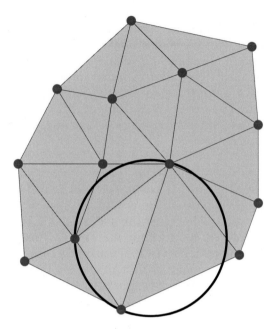

**FIGURE 12.5**
Delaunay triangulation.

The conversion from a raster to a TIN requires the determination of a $Z$ value tolerance, which represents the vertical accuracy. Ancillary data may be needed to improve the surface definition. Raster to TIN first generates a candidate TIN using sufficient input raster points (cell centers) to fully cover the perimeter of the raster surface. Then, by reiteratively adding more cell centers as needed, the TIN surface is incrementally improved until it meets the specified $Z$ tolerance.

## 12.3 Surface Data Analysis

Based on the data models constructed, 3D-based analysis can be conducted to solve real-world problems, such as calculating the elevation of a certain place, presenting slope and aspect of surface terrain, and deriving hydrological flow.

### 12.3.1 Elevation

Elevation of a certain point can be calculated based on the interpolation methods introduced in creating the surface. In a Grid model, elevation of a certain point can be calculated using the Grid cells close to the point. Taking IDW as an example, a general form of finding an interpolated elevation $z$ at a given point $x$ based on samples $z_i = z(x_i)$ for $i = 1, 2, \ldots, N$ using IDW is an interpolating function (Equation 12.3).

$$z(x) = \begin{cases} \dfrac{\displaystyle\sum_{i=1}^{N} w_i(x)z_i}{\displaystyle\sum_{i=1}^{N} w_i(x)}, & \text{if } d(x, x_i) \neq 0 \text{ for all } i \\[4pt] z_i, & \text{if } d(x, x_i) = 0 \text{ for some } i \end{cases} \tag{12.3}$$

where $w_i(x) = 1/d(x, x_i)^p$ is a simple IDW weighting function (Shepard 1968), $x$ denotes an interpolated (arbitrary) point, $x_i$ is an interpolating (known) point, $d$ is a given distance (metric operator) from the known point $x_i$ to the unknown point $x$, $N$ is the total number of known points used in interpolation, and $p$ is a positive real number, called the power parameter.

In the TIN model, the three points associated with the triangle inside which the point falls, as well as other nearest points, can be used for calculation.

## 12.3.2 Slope

Slope identifies the steepest downhill slope for a location on a surface. Slope is calculated for each triangle in TINs and for each cell in raster. TIN is the maximum rate of change in elevation across each triangle, and the output polygon feature class contains polygons that classify an input TIN by slope. For raster, slope is determined as the greatest of the differences in elevation between a cell and each of its eight neighbors, and the output is a raster.

The slope is the angle of inclination between the surface and a horizontal plane, and may be expressed in degrees or percent. Slope in degrees is given by calculating the arctangent of the ratio of the change in height ($dZ$) to the change in horizontal distance ($dS$) (Equation 12.4).

$$\text{Slope} = \text{atan}(dZ/dS) \tag{12.4}$$

Percent slope is equal to the change in height divided by the change in horizontal distance multiplied by 100 (Equation 12.5).

$$\text{Percent slope} = (dZ/dS) \times 100 \tag{12.5}$$

For a raster slope, the values of the center cell and its eight neighbors determine the horizontal and vertical deltas. As shown in Figure 12.6, the neighbors are identified as letters from a to i, with e representing the cell for which the aspect is being calculated.

| a | b | c |
|---|---|---|
| d | e | f |
| g | h | i |

**FIGURE 12.6**
Surface scanning window.

The rate of change in the $x$ direction for cell $e$ is calculated with Equation 12.6.

$$[dz/dx] = ((c + 2f + i) - (a + 2d + g)/(8 * x\_cellsize) \tag{12.6}$$

The rate of change in the $y$ direction for cell $e$ is calculated with Equation 12.7.

$$[dz/dy] = ((g + 2h + i) - (a + 2b + c))/(8 * y\_cellsize) \tag{12.7}$$

Based on the above Equations 12.4 through 12.7, the summarized algorithm used to calculate the slope is demonstrated in Equation 12.8.

$$\text{Slope}_{\text{radians}} = \operatorname{atan}\left(\sqrt{\left(\frac{dz}{dx}\right)^2 + \left(\frac{dz}{dy}\right)^2}\right) \tag{12.8}$$

Slope can also be measured in units of degrees, which uses Equation 12.9.

$$\text{Slope}_{\text{degrees}} = \operatorname{atan}\left(\sqrt{\left(\frac{dz}{dx}\right)^2 + \left(\frac{dz}{dy}\right)^2}\right) \times 57.29578 \tag{12.9}$$

### 12.3.3 Aspect

Aspect is the direction that a slope faces. It identifies the steepest downslope direction at a location on a surface. It can be thought of as slope direction or the compass direction a hill faces. Aspect is calculated for each triangle in TINs and for each cell in raster. Figure 12.7 shows an example of the aspect results of a surface using ArcMap 3D Analytics.

Aspect is measured clockwise in degrees from 0 (due north) to 360 (again due north, coming full circle). The value of each cell in an aspect grid indicates the direction in which the cell's slope faces (Figure 12.7).

Aspect is calculated using a moving $3 \times 3$ window visiting each cell in the input raster. For each cell in the center of the window (Figure 12.6), an

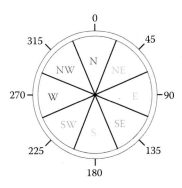

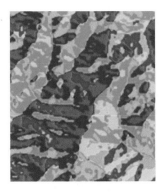

**FIGURE 12.7**
Clockwise in calculation aspect.

aspect value is calculated using an algorithm that incorporates the values of the cell's eight neighbors. The cells are identified as letters *a* through *i*, with *e* representing the cell for which the aspect is being calculated.

The rates of change in the *x* and *y* directions for cell *e* are calculated with Equation 12.10.

$$\begin{cases} [dz/dx] = ((c + 2f + i) - (a + 2d + g))/8 \\ [dz/dy] = ((g + 2h + i) - (a + 2b + c))/8 \end{cases} \quad (12.10)$$

Taking the rate of change in both the *x* and *y* directions for cell *e*, aspect is calculated using Equation 12.11.

$$\text{Aspect} = 57.29578 * \text{atan2}([dz/dy], \ -[dz/dx]) \quad (12.11)$$

The aspect value is then converted to compass direction values (0–360 degrees), according to the following rule:

if *aspect* < 0
    *cell* = 90.0 − *aspect*
else if *aspect* > 90.0
    *cell* = 360.0 − *aspect* + 90.0
else
    *cell* = 90.0 − *aspect*

Similarly, the aspect, or direction of the steepest downhill slope, can be calculated for each triangle in a TIN dataset and output as a polygon feature class. Each surface triangle's aspect is determined in units of degrees, then assigned an aspect code based on the cardinal or ordinal direction of its slope. Figure 12.8 shows the aspect calculation in ArcMap 3D Analyst.

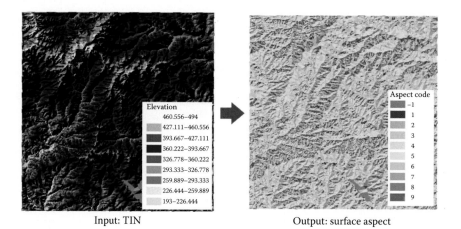

Input: TIN                    Output: surface aspect

**FIGURE 12.8**
Surface aspect (3D Analyst).

### 12.3.4 Hydrologic Analysis

One of the keys to deriving the hydrologic characteristics of a surface is the ability to determine the flow direction. In Grid, flow direction should be determined based on the grid cell and its eight neighbors. The flow direction is determined by the direction of steepest descent, or maximum drop, from each cell. This is calculated as Equation 12.12.

$$maximum\_drop = change\_in\_Zvalue / distance * 100 \qquad (12.12)$$

For raster surface dataset, the distance is calculated between cell centers. Therefore, if the cell size is 1, the distance between two orthogonal cells is 1, and the distance between two diagonal cells is 1.414 (the square root of 2). If the maximum descent to several cells is the same, the neighborhood is enlarged until the steepest descent is found. When a direction of steepest descent is found, the output cell is coded with the value representing that direction (Figures 12.9 and 12.10). Taking the 3 by 3 square (in red rectangle) as an example, the center cell (row 3, column 3) has a value of 44, surrounded by 8 neighboring cells. The steepest descent can be found at southeastern cell, which has the largest change from the center cell. Since the steepest descent direction is found to be the southeast, based on the direction coding (Figure 12.9), the flow direction of the center cell is 2.

If all neighboring cells are higher than the processing cell or when two cells flow into each other, creating a two-cell loop, then the processing cell is a *sink*, whose flow direction cannot be assigned one of the eight valid values. Sinks in elevation data are most commonly due to errors in the data. These errors are often caused by sampling effects and the rounding of elevations to integer numbers.

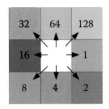

**FIGURE 12.9**
Direction coding.

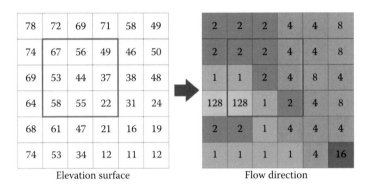

Elevation surface                    Flow direction

**FIGURE 12.10**
Flow direction example.

To create an accurate representation of flow direction, it is best to use a dataset that is free of sinks. Sinks should be filled to ensure proper delineation of basins and streams; otherwise, a derived drainage network may be discontinuous. The profile view of filling a sink is illustrated in Figure 12.11. For example, if the center pixel in Figure 12.9 is a sink, then the values of the nine pixels are sorted in a specified order (Figure 12.11) according to one of the several existing algorithms.

After filling sinks, flow direction can be conducted a second time to ensure accuracy. Flow accumulation can be determined based on the flow direction results. Accumulation is calculated according to the total number of cells that drain into a given cell and the value of each cell. First, the incremental flow (Figure 12.12b) of a given cell can be calculated by adding up the number of cells flowing into it. The total flow (Figure 12.12c) of a cell can then be calculated by summing the incremental flow value from the given cell and incremental flow values from all incoming cells.

Taking, as an example, the cell located at row 3, column 3 in Figure 12.12 (red rectangles), there are three arrows pointing to this cell, from northwest, west, and southwest, so that in the incremental flow, the value of this cell is 3. The accumulation value of this cell is obtained by adding this incremental value and the values from all the cells that flow into the cell. In this case, the calculation of accumulation flow is to add the incremental value 3 and 1

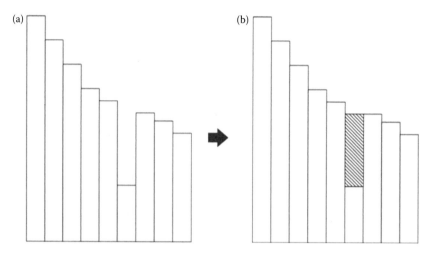

**FIGURE 12.11**
Profile view of a sink before and after fill.

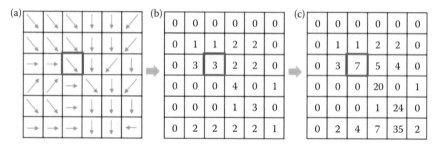

**FIGURE 12.12**
Flow accumulation calculations. (a) Flow directions. (b) Incremental flow. (c) Total flow.

from northwest cell, 3 from west cell, and 0 from southwest. Therefore, the accumulation flow of this cell is $3 + 1 + 3 + 0 = 7$.

## 12.4  Hands-On Experience with ArcGIS

### 12.4.1  Hands-On Practice 12.1: Conversion among DEM, TIN, and Contours

Step 1. Run Code 12.1 in the ArcMap Python window to create a TIN from a DEM surface raster dataset, and then generate a DEM from the TIN (Figure 12.13).

```
arcpy.env.workspace = r"C:\\ArcGISdata\\chp12data.gdb"

"""
    DEM to TIN
    The TIN data cannot be saved in a geodatabase, so the output data
    should be put into a folder e.g. C:\\ArcGISdata\\chp12data.gdb\\tin
"""
arcpy.RasterTin_3d("dem", r"E:\\ArcGISdata\\chp12data.gdb\\tin")

"""
    TIN to DEM
    The cell size of the new DEM is 50 meters, values are in the
    float type, and the methos used to raster the DEM is linear
    interpolation
"""
arcpy.TinRaster_3d(in_tin=r"E:\\ArcGISdata\\chp12data.gdb\\tin",
out_raster="demFromTIN",
                    data_type="FLOAT", method="LINEAR",
                    sample_distance="CELLSIZE 50",z_factor="1")
```

**CODE 12.1**
Conversion between DEM and TIN.

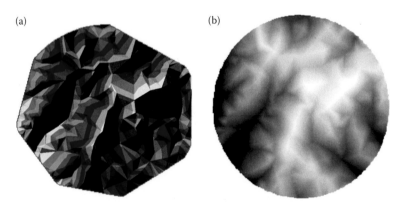

(a)  (b)

**FIGURE 12.13**
Result of Code 12.1. (a) TIN created from DEM. (b) DEM created from the TIN.

Step 2. Compare the original DEM and the new DEM that was regenerated from the TIN. Observe the areas with dramatic difference and
consider the reasons (Figure 12.14).

Step 3. Run Code 12.2 in ArcMap Python window to create contours
through a DEM surface raster dataset (Figure 12.15).

Step 4. Run Code 12.2 again, using the new DEM generated from the
TIN as input, to create another contour layer. Compare the two contour layers (Figure 12.16).

(a)　　　　　　　　　　　(b)

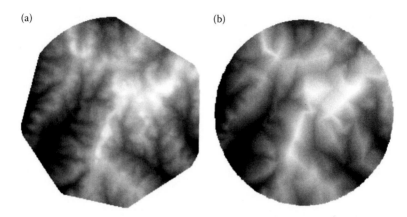

**FIGURE 12.14**
DEM comparison. (a) Original DEM. (b) DEM created from TIN, which is generated from the original DEM.

```
"""
    The input is "dem" and the output is "contour".
    The contour is in 10 meter intervals and starts from 330 meters.
"""
arcpy.Contour_3d(in_raster="dem", out_polyline_features=" contour",
            contour_interval="10", base_contour="330", z_
            factor="1")
```

**CODE 12.2**
Create contour from DEM.

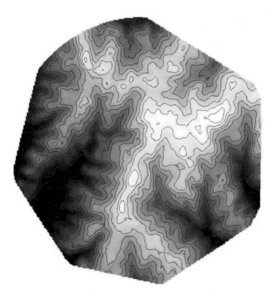

**FIGURE 12.15**
Result of Code 12.2—the contour created from DEM.

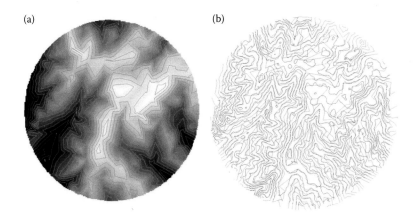

(a)  (b)

**FIGURE 12.16**
Contour comparison. (a) Contour generated from the new DEM. (b) Pink line is the contour generated from original DEM and green line is the one from the new DEM.

## 12.4.2 Hands-On Practice 12.2: Generate Slope and Aspect

1. Run Code 12.3 in ArcMap Python window to create slope from DEM (Figure 12.17).
2. Run Code 12.4 in ArcMap Python window to create aspect from DEM (Figure 12.18).

## 12.4.3 Hands-On Practice 12.3: Flow Direction

1. Run Code 12.5 in the ArcMap Python window to create flow direction matrix from DEM (Figure 12.19).
2. Continue to run Code 12.6 in the ArcMap Python window to check whether there are any sinks (i.e., cells without outlet). If any sinks exist, fill them on the original DEM to create a new DEM that will not generate any sinks upon flow direction calculations.
3. Continue to run Code 12.7 in the ArcMap Python window to create a flow direction and then a flow accumulation layer using the "no-sinks" DEM you created. Water streams will be observed in the flow accumulation layer (Figure 12.20).

```
# set the workspace
arcpy.env.workspace = r'C:\\ArcGISdata\\chp12data.gdb'

# the input is DEM, and slope is in the unit of degrees
slopely = arcpy.sa.Slope("dem", "DEGREE")
# save the slope layer into geodatabase (path has been set above)
slopely.save("slope")
```

CODE 12.3
Create slope from DEM.

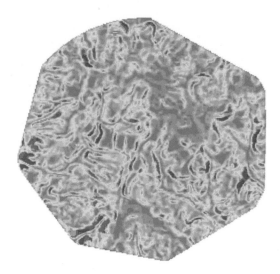

**FIGURE 12.17**
Result of Code 12.3.

```
# set the workspace
arcpy.env.workspace = r'C:\\ArcGISdata\\chp12data.gdb'
aspectly = arcpy.sa.Aspect("dem")
aspectly.save("aspect")
```

**CODE 12.4**
Create aspect from DEM.

**FIGURE 12.18**
Result of Code 12.4.

```
"""
    Set the workspace.  All new raster layers generated will be stored
    in the workspace.
"""
arcpy.env.workspace = r'C:\\ArcGISdata\\chp12data.gdb'

"""
    Create flow direction with "dem" as input.  The "NORMAL" argument means
    edge cells are not forced outward, but follow normal flow rules.
"""
fd = arcpy.sa.FlowDirection("dem","NORMAL")
```

**CODE 12.5**
Create flow direction from DEM.

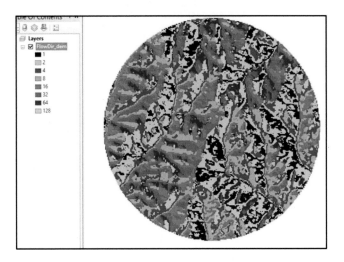

**FIGURE 12.19**
Result of Code 12.5.

```
# calculate sink
sinks = arcpy.sa.Sink("fd")
# fill the sinks on dem
dem_sinkfilled = arcpy.sa.Fill("dem")
```

**CODE 12.6**
Check and fill sink.

```
# recreate flow direction on the dem with sinks filled
fd_filled = arcpy.sa.FlowDirection("dem_sinkfilled","NORMAL")
# calculate the flow accumulation
fa = arcpy.sa.FlowAccumulation("fd_filled","","INTEGER")
```

**CODE 12.7**
Recreate flow direction and calculate flow accumulation.

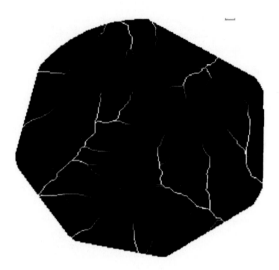

**FIGURE 12.20**
Flow accumulation layer generated by running Codes 12.3 through 12.7.

## 12.5 Chapter Summary

This chapter introduces the fundamentals of 3D surface data and the related basic processing algorithms and demonstrates how to program them in ArcGIS using Arcpy. Contents include the following:

1. Surface data structure and essential representations of surface data, that is, discrete data or continuous surface data.
2. The creation of two types of continuous surface data: Grid and TIN, and the conversion between them.
3. Surface data analysis, including the calculation of elevation, slope, aspect, and flow direction.
4. Hands-on experience with arcpy, conducting conversion among DEM, TIN, and contours, and calculating slope, aspect, and flow direction.

### PROBLEMS

1. Review the chapter and 3D Analyst extension of ArcGIS.
2. Pick a problem related to 3D surface analysis.
3. Design a solution for the problem, which should include the transformation to TIN and Grid, slope analysis, and aspect or flow direction analysis.

4. Select related 3D datasets to evaluate.

5. Code in Python to program a tool added to the ArcToolBox for operating ArcGIS and 3D Analyst and use it to obtain the results data/information needed in your solution.

NOTE: All codes can be successfully executed on ArcGIS for desktop versions 10.2.2 through 10.3. There may be a problem on running the code on the more recent version of ArcGIS.

# Section IV

# Advanced Topics

# 13

## Performance-Improving Techniques

Performance is critical in developing a GIS tool to accomplish tasks through time (Gittings et al. 1993). For example, a program that may need one hour to process line intersection can be improved to a few seconds for supporting near-real-time GIS tasks. This chapter discusses several programming performance improvement techniques (Yang et al. 2005) and strategies (Tu et al. 2004), including (a) fundamental computer engineering methods, such as accessing a file on an external storage and output device, (b) computational techniques, such as using parallel processing to speed up the performance, and (c) spatial methods, such as spatial index for improving access to large amounts of GIS data. General ideas, exemplar algorithms, and programming demonstrations are provided in this chapter. The fundamental intersection algorithm is used to illustrate how these techniques can be utilized to improve performance.

## 13.1 Problems

If the waiting time for processing cannot be tolerated by the end user, then the performance of a tool or computer software is critical to its success. Many GIS algorithms are time-consuming. For example, given the 10,000 GIS river features of the United States, finding the features within Washington, D.C., can be accomplished by evaluating each river to see whether or not it intersects with the D.C. boundary. Supposing that one such evaluation takes 0.1 second, the entire process could take $0.1 \times 10,000 = 1000$ seconds, or approximately 18 minutes. Such a long time is not tolerable to end users and is not typical to commercial software or optimized open-source software. As another example, if we have 10,000 roads and 10,000 rivers within the United States, and we want to find all intersecting points, we need to check for possible intersections between each road and each river. If such an average intersecting check takes 0.01 second, the entire process would take $10,000 \times 10,000 \times 0.01$ seconds = 1M seconds or approximately 300 hours, which is not tolerable. Many techniques can be utilized to improve processing. We will take a closer look at three different techniques to examine how and to what extent they can improve performance, as well as their limitations and potential problems.

Returning to the second example, calculate the line intersection by

(a) Reading both data files into memory
(b) Constructing polyline features for both datasets features
(c) Looping through each feature of both river and road data to conduct polyline intersection check
(d) Looping through each line segment of the two polylines to check whether they intersect
(e) Keeping all the intersection points
(f) Finalizing and visualizing or writing into a file, the results

The computing performance of the above steps can be improved in different ways as detailed in Sections 13.2 through 13.4. Step (a) can be improved with methods described in Section 13.2 by reading all data into memory to save the time for going back to the hard drive every time we fetch binary data for further processing. Steps (b) through (e) can be improved by two spatial methods (Section 13.4), such as using bounding box to filter out those not possible or using spatial index to go directly to the ones that have the intersection potential, and parallel methods (Healey et al. 1997) of executing the calculations using multithreading or grid, Hadoop cluster (Aji et al. 2013), and cloud computing (Yang et al. 2011a) techniques (Section 13.3). Steps (f) and (g) can be improved using multithreading (Section 13.3). Multithreading techniques (Section 13.3) can also be adopted to improve performance in steps (a) and (b).

The rationale behind the methodology described above is as follows: (1) accessing data through input/output devices (e.g., hard drive and monitor) is much slower than accessing data from RAM; (2) some calculations against static data and spatial relationship, such as line intersection, can be executed concurrently because each intersection calculation is independent from other intersection calculations. This means the line intersection results of line n and line m will not impact the line intersection between line i and line j; (3) spatial data are spatially constrained. For example, if the bounding boxes of two lines do not intersect, then the two lines cannot intersect. You can, therefore, filter out many unnecessary calculations by first employing the relatively simpler evaluation of bounding box intersections, as defined by their four coordinates (minx, miny, maxx, maxy).

## 13.2 Disk Access and Memory Management

In a computer system, the components are integrated through several approaches: the tightly coupled components (such as cache and register) are integrated inside the CPU, and the computer motherboard data bus integrates the CPU with other on-board components, such as RAM and others built on the motherboard. Many other components, such as printers, massive storage devices, and networks, are connected through an extension (e.g., a cable), from the motherboard. Hard drive (HDD) and massive storage are among the most frequently used external devices to maintain data/system files. Recent advances in computer engineering have enabled more tightly coupled storage on the motherboard using bigger RAM and ROM sizes. In general, a hard drive is slower than RAM in terms of data access; therefore, we can speed up a file access process by reading only once from HDD into RAM and then operating in the memory multiple times instead of reaccessing HDD many times (Yang et al. 2011c).

### 13.2.1 File Management

In Python, file access is processed in three steps: open file, access file, and close file. The first and third steps are common for file access and management. The access file step can be quite different depending on how you read from or write data to the file.

As a specific example, consider accessing point shapefile data: this process includes reading the file header information to obtain the number of points and the bounding box for the point shapefile. The rest of the process involves rotating through the read point data process to get the x, y coordinates for all points, and creating a feature point to add to the point data layer. Code 13.1a shows reading from the hard drive once for all data, and then processing the RAM-based data element by element to retrieve the information. Code 13.1b shows reading from the hard drive every time a point feature is retrieved. Changing the FileBuffer variable to true or false will switch between these two reading modes at the ReadShapeFile.py module in the performance package shipped with this book.

By switching the FileBuffer between true and false for reading one layer of Amtrack station data (amtk_sta.shp) on a laptop, we obtained a value of ~0.060 seconds for true and a value of ~0.069 seconds for false. These values may change according to different datasets, computers, and working loads of the computers.

### 13.2.2 Comprehensive Consideration

The specific improvements will differ greatly according to the specific configurations on the computer hardware (Peng 1999). For example, if the file size exceeds the size of the RAM, you cannot use this method, and you will need to device a different strategy, for example, reading the file in chunks instead of reading the entire file. The difference between the speed of RAM and speed of HDD should also be considered. The faster the RAM and slower the HDD of a computer, the better speed improvement you can obtain using this a:

```
if bFileBuffer:
        size = os.path.getsize(fileName)
        shpFile=open(fileName,'rb')
        s = shpFile.read(size)
        shpFile.close()
        b = struct.unpack('>i',s[24:28])
        b=b[0]*2
        featNum = (b-100)/28
        shpFile.close()
        layer.minx, layer.miny, layer.maxx, layer.maxy = struct.
        unpack("<dddd",s[36:68])
        pointer = 100+12
        for i in range(0,featNum):
            b = struct.unpack('dd',s[pointer:pointer+16])
            point = FTPoint(b[0],b[1])
            layer.features.append(point)
            pointer+=28
```

b:

```
    else:
        shpFile=open(fileName,'rb')
        s = shpFile.seek(24)
        s = shpFile.read(4)
        b = struct.unpack('>i',s)
        b=b[0]*2
        featNum = (b-100)/28
        s = shpFile.read(72)
        header = struct.unpack("<iiddddddddd",s)
        layer.minx, layer.miny, layer.maxx, layer.maxy =
        header[2],header[3],header[4],header[5]
        for i in range(0,featNum):
            shpFile.seek(100+12+i*28)
            s = shpFile.read(16)
            b = struct.unpack('dd',s)
            point = FTPoint(b[0],b[1])
            layer.features.append(point)
        shpFile.close()
```

**CODE 13.1**

Reading point data from a shapefile with using a single- or multiple-access process: Code (a) reads all data from the file at once using shpFile.read (size), and unpacks the data from the content read to s, which is kept as a variable in memory. Code (b) reads data from the hard drive by jumping through the elements needed and unpacking the data while moving ahead.

method. A laptop will have a less drastic speedup than a desktop using this method, because the speed difference between accessing RAM and HDD on a desktop is greater than that of a laptop.

The code for both polyline and polygon shapefiles have been added in the performance.zip package. Different data layers can be added to the main. py file and run it to test the time needed for reading the shapefiles with and without the buffer option.

The more external devices used, the slower the program. For example, if "print" is added to the FTPoint class initialization function def __init__(self), the process will become much slower if the same main file runs without changing anything else.

## 13.3 Parallel Processing and Multithreading

Progressively building knowledge in sequential order means that later parts depend on the earlier parts. For example, you have read about GIS programming from Chapters 1 through 13 in sequential order. In computer processing, if the process includes many steps and each step depends on the previous step, then the execution will take much longer. But if the steps are not dependent on the previous steps, then the steps can be executed in parallel (Healey et al. 1997). For example, the three layers of data (Amtrack station, states, and national highway lines) must be loaded for making the initial map. If they are not dependent on each other for processing, then they can be read and executed in three parallel processes.

### 13.3.1 Sequential and Concurrent Execution

The "Programming Thinking" chapter (Chapter 5) discusses how to analyze an application problem in sequence so that there is a clear logic about the start, process, and end of an application. In a sequential execution process, the statements written for an application will be run one by one and in the sequence defined by the program logic. It is like one person taking the statement instructions and executing them one at a time. The computer components, however, are separated and linked through different data and command connections. This separation means that different components can be utilized to execute different processes at the same time. For example, if you read the hard drive to the RAM data file, you could also conduct mathematical calculations in the CPU at the same time. Modern computer architectures allow many processes to be executed concurrently. It is like when there are many people to do different tasks simultaneously, the entire application task can be accomplished earlier. Python programming language provides the multithreading (Tullsen et al. 1995) module to support such concurrent process.

### 13.3.2 Multithreading

The performance package integrates a sample multithreading framework. The logic workflow repeats a 0.01-second process 1000 times. Processed sequentially, this will require roughly 10 seconds to finish all processes.

```
a:

threads = []
for i in range(10):
    threads.append(SummingThread(i*100,(i+1)*100))

starttime = time.clock()

for i in range(10):
    threads[i].start() # This actually causes the thread to run

for i in range(10):
    threads[i].join() # This waits until the thread has completed
# At this point, both threads have completed
result = 0
for i in range(10):
    result+=threads[i].total
b:

thread = SummingThread(0,1000)
starttime = time.clock()
thread.start()
thread.join()
print 'single thread'
print thread.total
print str(time.clock()-starttime) + ' seconds\n'
```

**CODE 13.2**
Execute the same process (takes 0.01 second) 10,000 times in 10 multithreads (a) versus in one single thread (b).

However, you can break the required processes into 10 groups, with each group processed by a thread, so all 10 groups can be executed concurrently. Ideally, this approach could reduce the processing time from 10 seconds to 1 second. Code 13.2a creates a 10-thread list and Code 13.2b creates one thread and executes the same number of processes. The SummingThread is inherited from the Python module *threading* class *Thread*.

The code is included in the multithreading.py file and the execution will output the time spent by each method. It is observed that the 10-multithread approach is about 10 times faster than the single-thread approach by running the Python file. The code can be experimented on by running on one computer or multiple computers to compare the time outputs to observe the improvements.

### 13.3.3 Load Multiple Shapefiles Concurrently Using Multithreading

One example of such concurrent processing is loading shapefile data into RAM to create objects. The main Python file in the performance package includes a package to load three shapefile data files concurrently if the multithreading variable is set to *True*. If it is *False*, then the three shapefiles will be loaded sequentially. Code 13.3 illustrates the utilization of multithreading

```
a:

if multithreading:
    starttime = time.clock()
    lr1 = AddMapLayer(map,'amtk_sta','yellow')
    lr2 = AddMapLayer(map,'amtk_sta', 'red')
    lr3 = AddMapLayer(map,'amtk_sta','pink')
    lr1.start()
    lr2.start()
    lr3.start()
    lr1.join()
    lr2.join()
    lr3.join()
    print (str(time.clock()-starttime) + ' seconds')

b:

else:
    starttime = time.clock()
    map.addLayer('amtk_sta','yellow')
    map.addLayer('amtk_sta', 'blue')
    map.addLayer('amtk_sta','red')
    print (str(time.clock()-starttime) + ' seconds')
```

**CODE 13.3**
Loading three data layers concurrently (a) or in sequence (b).

to read data (a) versus reading data sequentially (b). The AddMapLayer class is a multithreading class defined as based on the Thread class of threading module.

The main Python file can be experimented on by running on a computer and switching the multithreading Boolean variable on or off and recording the time spent on each approach. As an exercise, run the code 5 times with each approach and record the average values. Also try this on different computers or compare your results with peers if they used different computers.

## 13.3.4 Parallel Processing and Cluster, Grid, and Cloud Computing

Executing tasks concurrently (i.e., processing in parallel) will help speed up the entire application; however, there are exceptions to its use. If the application is executed in sequence and each statement depends on the previous one, then the process cannot be parallelized. Different computers will have different capacities for handling different numbers of concurrent threads based on the computer structure and its current usage. For example, an 8-core CPU desktop will be able to handle more concurrent computing threads than a single-core CPU desktop. The applicability of multithreading is determined by the parallelizability of the process and the maximum number of concurrent processes the computer architecture can handle. For example, does it have multiple CPU cores and does it support calculation in GPU cores? The program will also impact how many multithreads can be executed concurrently.

The example in Section 13.3.3 reads multiple shapefiles. If the file buffer strategy is used, as discussed in Section 13.3.1, then it is possible to execute more concurrent threads because when several threads access intensively, the hard drive will let the threads compete for the same I/O resource, thereby degrading the performance gain. The final and actual performance gain can be determined by testing a combination of different computing techniques.

Many twenty-first century challenges require GIS data, processes, and applications, and users are widely distributed across the globe (Ostrom et al. 1999). For example, when a tsunami hits the Indian Ocean, people along the coast are impacted and need GIS to guide them in making safe decisions. An application built on sequential strategy will not be able to handle this problem in a way that satisfies end users or provides timely decision support information. Therefore, supercomputers are adopted to provide information in near real time by processing the data much more rapidly using closely coupled CPUs and computers (Zhang 2010). Grid and cloud computing infrastructure are utilized to share data, information, and processing among users (Yang et al. 2013). Using such infrastructure would help improve the application performance. But the detailed improvements have to be tested and optimized to gain the best performance in a computing infrastructure. In Xie et al. (2010), a high performance computing (HPC) example is illustrated, which has a dust model parallelized and running on an HPC environment to speed up the process. Results show that with the initial increase of CPU/server numbers participating in the geospatial simulation, the performance is rapidly improving. The time spent is reduced by nearly half by increasing from one core to two cores, and from two cores to four cores; however, increasing beyond eight CPU cores will not further increase performance. Although concurrent processing can help increase processing speed, the dust simulation is a spatiotemporal phenomenon, meaning that the dust concentration moves across different simulation subdomains to maintain its natural continuity. This process of sharing dust data among different subdomains, known as model synchronization, is critical. The running time may actually increase if the domain is divided among subdomains on different computers. The benefit of running in parallel and the cost of synchronizing is observed to reach a balance at a certain point—in this case, when using eight CPU cores.

## 13.4 Relationship Calculation and Spatial Index

Most spatial relationship calculations are complex and require examining each feature within a data layer (Dale et al. 2002). The problem introduced at the beginning of Section 13.1 would take a long time if you were to find polyline intersections by executing the entire calculation process for all data features. This is not acceptable in GIS software and tools. Fortunately, using spatial principles will optimize the process by filtering out complex calculations. An important component is the feature bounding box, defined by the four coordinates of minx, miny, maxx, and maxy for a minimized rectangle enclosing a feature. The spatial pattern or principle is that when the bounding boxes of two features disjoint from each other, the two features must be disjointed. If you are to calculate the intersection of a river data layer and road data layer (Section 13.1), a bounding box of a river in Washington, DC will not intersect with the bounding box of roads in California. This spatial pattern can be utilized to filter out most features before starting the complex computation of the intersection as introduced in Chapter 8. Use the data layer intersection as an example to introduce how to build the algorithm into MiniGIS and how to optimize the algorithm using a bounding box in Section 13.4.1. Another simplified spatial pattern, applied to one-dimensional data, is to sort features (such as points) according to a spatial dimension and to conduct filtering according to a tree structure, such as binary tree. By expanding this idea to two dimensions, many different types of spatial indices can be built to speed up spatial relationship calculations (detailed in Section 13.4.2).

### 13.4.1 Bounding Box in GIS

Bounding box is widely used in GIS (Agarwal et al. 2012). For example, in shapefile, the entire shapefile bounding box is maintained in both the .shx and .shp files. Each polyline and polygon's bounding box is also maintained in .shp files to facilitate the usage of bounding box. As an exercise, find the intersection points of two data layers: rivers and roads. These datasets are kept in the map object of the MiniGIS package. This section will introduce the logic related to implementing the intersection algorithm based on the map object with two data layers. The detailed implementation is explained in the hands-on experience section (Section 13.5).

The map object includes access to all other objects related to GIS data and map operations of the mini-GIS packages. From the programming thinking perspective, try to identify the intersection points of any river and any road kept in the map object. In order to find this, we need to repeat each river and each road to check whether they intersect with each other. Each river or road is a polyline feature composed of multipart lines that include many points. Therefore, we need to repeat calculations for every river line segment and every road line segment to identify the potential intersections. Each line segment intersection is identified through the lineseg class defined in Chapter 8. This process can, therefore, be interpreted as the following workflow:

1. Check the intersection of two data layers in the map (can be selected from existing layers in GUI).
2. Check the intersection of two features in the data layers, using one from each layer (rivers and roads), and keep all intersecting points (Layer class).

3. Check the intersection of every line segment pair, using one from each layer (rivers and roads), and keep all intersecting points (Polyline class).

4. Check the intersection and calculate the intersecting point (LineSeg class).

To speed up the process, add the bounding box check to the first three steps to:

1. Check whether two data layers are intersecting with each other and return false if not.

2. Check whether two features' bounding boxes are intersecting with each other and return false if not.

3. Check whether two line segments' bounding boxes intersect with each other and return false if not.

The bounding boxes' checking algorithm is relatively simple for bounding box one (bb1) and bounding box two (bb2):

bb1.minx > bb2.maxx or bb1.maxx<bb2.minx or bb1.miny>bb2.maxy or bb1.maxy<bb2.miny

If this statement is true, then the two bounding boxes cannot intersect with each other. If this is not true, the two bounding boxes intersect with each other and we can proceed with the following, more time-consuming, calculations.

### 13.4.2 Spatial Index

In the previous bounding box-based improvement, all features and all line segments were checked to test for the condition of bounding box intersection. This is practical when the number of features is small. But when the number of polylines or line segments is massive, this iterative process will take a lot of time. Given a target bounding box, another method is to look for features that only fill in bounding boxes intersected with the target one. This can be achieved by using the spatial index (Samet 1990), which is built based on spatial patterns and principles. This section introduces the popular *R-Tree index* (Guttman 1984) as a spatial index example.

R-Tree is a very popular spatial index based on the bounding box (rectangle) of polylines and polygons. The general principle of R-Tree is to build a tree with features close to each other inside a branch of the tree. When searching through an R-Tree, if the upper-level branches' bounding box does not intersect with a query rectangle/polygon, then it is not possible that features from the branch will intersect with the query shape. Figure 13.1 shows an R-Tree index with eight polygonal objects using their bounding box. The entire domain is divided into two big rectangles with some overlap. Each rectangle includes

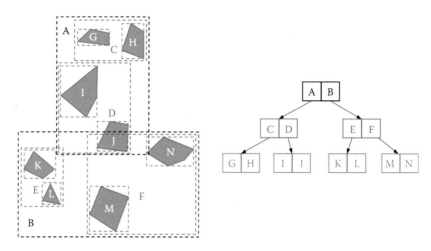

**FIGURE 13.1**
R-Tree example.

four polygons. Each rectangle is further divided into two smaller rectangles that are, again, divided into smaller rectangles, each containing one polygon. In dividing, three types of criteria were applied: (a) the closer polygons are put in the same rectangle, (b) each rectangle is the minimum rectangle that includes the bounding boxes of all polygons within that rectangle, and (c) the process of division yields two rectangles from the original one. These criteria ensure that the closer polygons are inside the same branch of the R-Tree and that the R-Tree is a balanced binary tree, which provide better performance. There could be other criteria applied in the division process, such as minimizing the overlap of rectangles from the same level of division. The left side of the figure shows mapping polygons and R-Tree division; the right side shows the resulting R-Tree. When searching for features, only the branch that intersects with the searching rectangle will be further searched. This will greatly improve the performance when there are many features included.

## 13.5 Hands-On Experience with Mini-GIS

The techniques are implemented in map, polyline, lineseg, and layer classes and called from the main program. Switches are provided to turn on and turn off the method. The four datasets are provided to test the performance-improving techniques by flipping the switches.

### 13.5.1 Data Loading with RAM as File Buffer

Hard drive operations are much more time-consuming than RAM access. Different computers have different performance results. This hands-on experience is used to test the effectiveness of using RAM as a file buffer when reading from the hard drive. Use RAM to hold all file content read in, and process from RAM instead of accessing hard drive for each file operation. This method is implemented in the ReadShapeFile.py file.

Hands-on experience on file buffer:

1. Open the main.py and ReadShapeFile.py files.
2. Run the main.py file and record the time spent on loading data.
3. Run (2) 5 times after the map window is closed for 10 seconds for each time.
4. Flip the bFileBuffer variable.
5. Repeat (2) and (3) and record the time, then compare the time spent on each method and analyze the result.
6. Change computers (e.g., from desktop to laptop or from laptop to desktop) and repeat steps (1) through (5).
7. Analyze the time differences obtained among different computers.

### 13.5.2 Data Loading with Multithreading

Many processes in a GIS application involve parallelizable steps, and can, therefore, be sped up using multithreading methods. One such process is to load multi-data layers for the initial map as detailed in Section 13.3.2. You can do a sample test to find the efficiency of this method by turning on and off the multithreading in the main.py file, which added the multithreading for opening and adding data layers.

Hands-on experience on multithreading:

1. Open the main.py file.
2. Run the program to record how much time is spent with multithreading switched to off.
3. Run this for 10 times and record the processing times.
4. Switch the multithreading to on and repeat step (3).
5. Compare the processing times and analyze the result.
6. Replace the three datasets and repeat steps (3) through (5).
7. Copy and paste the same data with different names and run the same experiments.
8. Switch to another computer and run the same experiments.
9. Analyze the patterns found in your recorded results.

### 13.5.3 Bounding Box Checking to Speed Up Intersection

Based on analyses of Section 13.4.1, we have four steps to find the intersecting points as (a) check two data layers, (b) check two features, (c) check two line segments, and (d) the line segment intersection algorithm. We can add a bounding box check for the first three steps, which will reduce the number of time-consuming calculations of (d) and improve the program performance. The three bboxcheck functions are defined in the Layer, Polyline, and LineSeg classes (Code 13.4). Because each layer has many polylines and each polyline has many line segments, this process could significantly improve the performance. Code 13.4 illustrates how the intersect and bboxcheck functions are defined in Layer class. The Layer's bboxcheck is called by Map class. The Layer's intersect function calls the Feature's bboxcheck to avoid calculating two features whose bounding boxes do not intersect. The Polyline class has similar definitions, with its bboxcheck defined by Layer class object, and its intersect function calling the bboxcheck function of LineSeg class objects.

Once the calculations are completed, you can display the resulting points on the GUI. This can be handled in several ways, including (a) adding a data layer to store the points as the "intersecting point layer," (b) maintaining a data structure, for example, a list of intersecting points of map object to keep all points, and (c) associating the points with one of the two layers. To simplify the process, choose the second method by maintaining a list data structure. After the intersection calculations, obtain a list of points defined as self.intersectPoints in the Map class of map.py file. When displaying the points after the calculation, you need to go through a process similar to what you used to display points on Tkinter (Code 13.5)

- Call self.vis() method to redisplay all data.
- Go through all data layers that are visible.
- Go through all points to transform the geographic coordinate system from original to window coordinate system.

```
def bboxcheck(self,layer):
        if self.minx>layer.maxx or self.miny>layer.maxy or self.
        maxx<layer.minx or self.maxy<layer.miny:
            return False
        else:
            return True
    def intersect(self,layer):
        intPoints = []
            for feature1 in self.features:
                for feature1 in self.features:
                    if feature1.bboxcheck(feature2) or
                    noBoundingboxCheck:
                        retPts = feature1.intersect(feature2)
                        if retPts:
                            for point in retPts:
                                intPoints.append(point)
        return intPoints
```

**CODE 13.4**
Using bounding box check to filter out most intersection calculations in Layer class.

```
def vis(self):
    self.can.delete('all')
    self.calculate()
    for layer in self.layers:
        for feature in layer.features:
            feature.vis(self, layer.color)
    for point in self.intersectPoints:
        xy = self.transform(point)
        self.can.create_rectangle(xy[0]-4, xy[1]-4, xy[0]+4,
        xy[1]+4, fill='brown')
    self.can.pack()

def transform(self, point):
    if (self.controlPoint == 1): #TOPLEFT
        winx = int((point.x-self.minx)*self.ratio)
        winy = int((self.maxy-point.y)*self.ratio)
    elif (self.controlPoint==2): #CENTER
        winx = int((point.x-(self.minx+self.maxx)/2)*self.
        ratio)+self.windowWidth/2
        winy = int(((self.maxy+self.miny)/2-point.y)*self.
        ratio)+self.windowHeight/2
    elif (self.controlPoint==3): #LOWERLEFT
        winx = int((point.x-self.minx)*self.ratio)
        winy = int((self.miny-point.y)*self.ratio)+self.
        windowHeight
    elif (self.controlPoint==4): #TOPRIGHT
        winx = int((point.x-self.maxx)*self.ratio)+self.
        windowWidth
        winy = int((self.maxy-point.y)*self.ratio)
    else: #LOWERRIGHT
        winx = int((point.x-self.maxx)*self.ratio)+self.
        windowWidth
        winy = int((self.miny-point.y)*self.ratio)+self.
        windowHeight
    return winx,winy
```

**CODE 13.5**
Visualize the intersecting point as rectangles when displaying the data layers.

- Go through all points to create a visible symbol, for example, a square, for representing the point (Figure 13.2).

Hands-on experience: Investigate the effectiveness of bounding box checking:

1. Open the main.py, map.py, layer.py, polyline.py, and lineseg.py in Python IDLE GUI.
2. Change the Boolean value of noBoundingboxCheck to False as highlighted in Code 13.4 from layer.py, and save it.
3. Run the program from main.py and click on the intersection button.
4. Record the time spent on checking intersections.

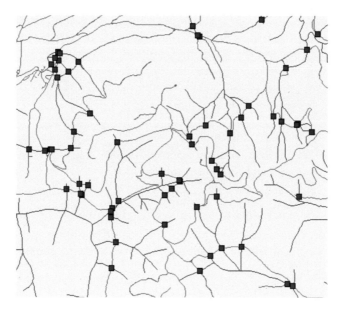

**FIGURE 13.2**
Visualization result of Code 13.5.

5. Change the Boolean value of noBoundingboxCheck to True as high-lighted in Code 13.4 from layer.py, and save it.

6. Run the program from main.py and click on the intersection button.

7. Record the time spent on checking intersections without bounding box checking for feature level.

8. Compare the 2 times obtained (we obtained 2 seconds and 122 sec-onds, respectively, with and without bounding box check at the fea-ture level called in layer.py).

9. Repeat this process to comment out "if (self.layers[1].bboxcheck(self. layers[2])):" from map.py and "if ls1.bboxcheck(ls2):" and record the time before and after commenting them out (we obtained 2 seconds vs. 2.2 seconds, respectively, with and without the line segment bounding box check).

10. Use different combinations of the three bounding box checks to find out which one is most important, and analyze the patterns found.

11. Change the data to do the same experiment and check how the results have been changed; analyze the patterns found.

### 13.5.4 Line Intersection Using R-Tree Index

The previous subsections compared the time needed for calculating line intersections with and without using bounding box. It is observed that the bounding box could significantly improve the performance of the program. This subsection illustrates how R-Tree index could help further improve the performance using the code embedded in the Layer.py file. The R-Tree packages from Python Software Foundation need to be installed before the testing (e.g., using pip to install Rtree-0.8.2-cp27-cp27m-win32.whl).

Hands-on experience on R-Tree index:

1. Open the Layer.py and main.py file.
2. Change the Boolean value of noBoundingboxCheck to True.
3. Change brtree = False in Layer.py file.
4. Run main.py file.
5. Zoom and pan to show the map within the full window.
6. Click on the intersection button and record the time spent on the intersection calculation. This step may take a few minutes.
7. Close the application.
8. Change brtree = True in Layer.py file.
9. Run main.py file.
10. Zoom and pan to show the map within the full window.
11. Click on the intersection button and record the time spent on the intersection calculation.
12. Repeat steps (2) through (10) 5 times.
13. Analyze the computing time patterns found in the experiment.

---

## 13.6 Chapter Summary

This chapter explores performance-improving techniques for GIS programming and includes

1. Performance challenges when dealing with big data or computing-intensive process.
2. Buffering techniques for reducing the number of accesses to slow devices, for example, hard drive.
3. Multithreading techniques for executing parallelizable processes concurrently.
4. Bounding box check for significantly reducing the number of time-consuming calculations.
5. Spatial index to get to datasets of interest directly instead of iterating through all data.

## PROBLEMS

The objective of this homework is to understand and design a comprehensive performance tuning and management process.

1. Please select four different datasets or use the four datasets provided.
2. Design a comprehensive performance testing experiment.
3. Conduct the tests using the MiniGIS package.
4. Compare the performance improvements before and after adopting the techniques.
5. Explain the performance differences and discuss the trade-off when using different techniques.

NOTES: The results may differ according to the datasets selected and the computers used to run the MiniGIS. The applicability of the three different categories of techniques will determine the final performance of the software.

# 14

## Advanced Topics

GIS algorithms and programming are critical to the research and development in advancing geographical information sciences (GIScience), because most mature GIS software packages are not flexible enough to be revised for the purpose of testing new ideas, models, and systems. This chapter introduces how GIS programming and algorithms are utilized in advancing several GIScience frontiers.

### 14.1 Spatial Data Structure

Spatial data structure is the logical and computational representation of spatial datasets extracted and modeled from real world for solving a group of application problems. Spatial data structure falls into two categories: (1) vector data, such as point, polyline, and polygon to represent cities, rivers, land surfaces, etc.; and (2) raster data as introduced in Chapter 10 to organize a matrix of cells into rows and columns where each cell has a value to represent information, for example, temperature, elevation, and humidity. A well-designed data structure should be able to provide accurate representation of the complicated data/problem and support efficient data access to obtain optimal performance for spatiotemporal query, access, and analytics. The netCDF/HDF-based raster data are taken as an example to show how it helps organize scientific raster data to support spatiotemporal data representation, access, and analytics. Based on the netCDF/HDF data, an advanced spatiotemporal index research is introduced.

#### 14.1.1 Raster Data Structure in NetCDF/HDF

The classic data model in netCDF and HDF consists of multidimensional variables within their coordinate systems and some of them are named auxiliary attributes (Rew and Davis 1990, Figure 14.1a). Each variable is specified by a list of named dimensions and other attributes. The dimensions may be shared among variables to indicate a common grid. Variables and attributes have six primitive data types: char, byte, short, int, float, or double. Given four-dimensional array-based datasets as an example (shown in Figure 14.2), each variable consists of three spatial dimensions (latitude, longitude, and

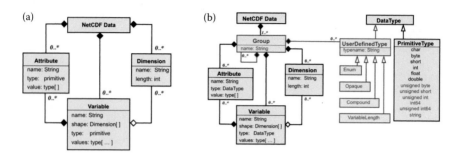

**FIGURE 14.1**
(a) The architecture of classic data model. (b) The architecture of the improved data model.
(From http://www.unidata.ucar.edu/software/netcdf/papers/nc4_conventions.html.)

altitude) and one temporal dimension (time). Each layer in the array has the same latitude and longitude dimension, and is called the "grid," which is used to store the values for each of the layer-specific variables. These grids are further grouped by altitude. The variables may have additional attributes to explain more properties, such as the variable's full name and the unit for the variable's value. However, the classic data model has two obvious limitations: (1) lack of support for nested structure, ragged arrays, unsigned data types, and user-defined types; and (2) limited scalability due to the flat name space for dimensions and variables.

To address the limitations of the classic data model, a new data model is proposed and implemented (Figure 14.1b). It adds a top-level, unnamed group to contain additional named variables, dimensions, attributes, groups, and types. Groups are like directories in a file system, each with its own set of named dimensions, variables, attributes, and types. Every file contains at least the root group. A group may also contain subgroups to organize hierarchical datasets. The variables depicted in Figure 14.2 can be divided into different groups by a certain characteristic, such as the model groups that generated these data. When storing these variables in a physical file, each 2D grid will be decomposed into one-dimensional byte stream and stored separately, one by one, in a data file.

### 14.1.2 Application of NetCDF/HDF on Climate Study

The Modern-Era Retrospective Analysis for Research and Applications (MERRA) is the latest reanalysis product by NASA's Global Modeling and Assimilation Office (GMAO) using the Goddard Earth Observing System, version 5 data assimilation system (GEOS-5) and providing data dating back to 1980. MERRA has been widely used to study weather and climate variability. These data are archived in the HDF-EOS format, based on HDF4. MAT1NXINT, one such MERRA product, and contains nearly 111 2D hourly variables with a spatial resolution of 2/3° longitude by 1/2° latitude. With

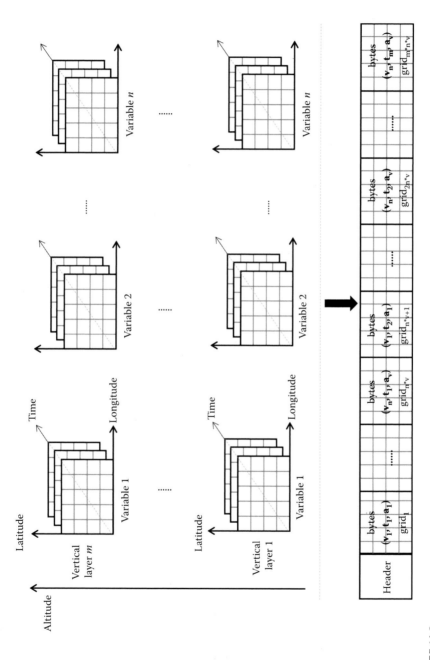

**FIGURE 14.2**
The logical structure of the improved data model.

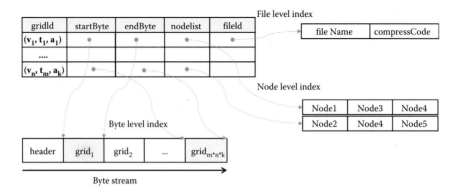

**FIGURE 14.3**
The structure of spatiotemporal index.

HDF4, all variables at a certain time point can be stored in a single file instead of multiple files.

To enable MERRA data, it can be analyzed in parallel without requiring preprocessing; Li et al. (2016) utilized the netCDF library for Java to extract the data structure information of MERRA files and build a spatiotemporal index, which is implemented as a hash index. Figure 14.3 depicts the structure of the spatiotemporal index. The index consists of five components: *gridId*, *startByte*, *endByte*, *fileId*, and *nodeList*. The first of these, *gridId*, records the variable name, time, and altitude information for each grid; *startByte* and *endByte* indicate the exact byte location of the grid in a file; *fileId* records the file location where the grid is stored; and *nodeList* records the node location where grids are physically stored in a Hadoop Distributed File System (HDFS). The key for the hash index is *gridId*, and the others are treated as the values. When querying the MERRA data, the spatiotemporal index will be traversed first. If the *gridId* has the same variable name as one of the queried variables, then the time and altitude information in *gridId* will be further compared with the input spatiotemporal boundary. If they are within the spatiotemporal boundary, the values (*startByte*, *endByte*, *nodeList*, and *fileId*) will be fetched out and utilized to read the real data out from physical disks using HDFS I/O API. The spatiotemporal index enables users to access data by reading only the specific data constrained by the input spatiotemporal boundary, eliminating the need to examine all of the data.

One month (January 2015) of the MAT1NXINT product (45.29 GB) was used as experimental data. MapReduce was adopted to compute the daily mean in parallel for a specified climate variable over a specified spatiotemporal range. Two scenarios were evaluated: the first scenario, as the baseline, was performed without using the spatiotemporal index; the second was performed using the spatiotemporal index. The experiments were conducted on a Hadoop cluster (version 2.6.0) consisting of seven computer nodes (one master node and six slave nodes) connected via 1 Gigabit Ethernet (Gbps).

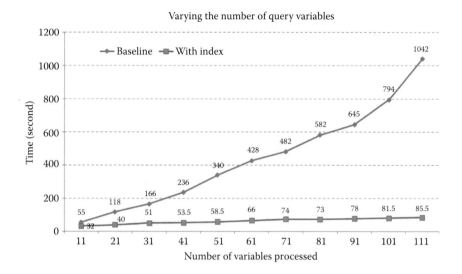

**FIGURE 14.4**
Run time for computing the daily global mean for January 2015 for different numbers of variables.

Each node was configured with eight CPU cores (2.35 GHz), 16 GB RAM, and CentOS 6.5.

Figure 14.4 shows the run times, comparing the baseline and the indexing approach for different numbers of variables. When increasing the number of variables in the query, the run time for the baseline condition increased from 55 to 1042 seconds—nearly 19 times longer. In contrast, when the spatiotemporal index was employed, the run time increased from 35 seconds to 85 seconds—only 2.4 times longer. This comparison result demonstrates that the spatiotemporal index can significantly improve the data access rate for the MERRA data stored in HDFS. Two factors lead to this improvement:

1. The spatiotemporal index can point exactly to data locations at the node, file, and byte level. It can tell MapReduce at which node the specified array of a variable is located, which file it is stored in, and its starting and ending bytes within the file. With this information, MapReduce can move the program close to the nodes where the referenced data are stored, minimizing data transfer by the network.

2. The data structure information is extracted and stored externally in the database. Therefore, when querying the data by variable names, only the database needs to be queried; there is no need to repeatedly access the metadata in MERRA data using HDFS I/O API, and, thus, avoiding the limitation of HDFS when randomly accessing small data.

## 14.2 GIS Algorithms and Modeling

GIS was developed to solve specific geospatial problems. In this process, a real-world problem is extracted into conceptual models, which are further abstracted into computable models. This process is called modeling, and the eventual result of the model is to produce proper information that can be used to support problem solving. For example, the weather forecasting model can take abstract and parameterized inputs into geographical calculations for simulating what weather might be expected at a future time stamp.

To demonstrate the GIS algorithms and modeling, the following example takes Twitter data from the 2015 Spring Festival in New York City and analyzes it for several phenomena of interest. The objectives are to identify the spatial distribution of tweets related to the Spring Festival, and to explore the socioeconomic variables that might explain this type of spatial distribution. Social media analysis is one of the hottest research topics in recent years, as it has become a novel avenue for the contribution and dissemination of information. Social media data from sources such as Twitter usually include references to when and where a given social media feed was created (Crooks et al. 2013). Through geosocial media, we are able, for the first time, to observe human activities in scales and resolutions that were previously unavailable. The Spring Festival, or Chinese New Year, is an important traditional Chinese festival celebrated on the Chinese lunar calendar. New York City was selected as the study case because, according to the 2010 Census, it is now home to more than one million Asian Americans, a number greater than the combined totals of San Francisco and Los Angeles.

### 14.2.1 Data

Three datasets are used in this example: tweets in NYC during Chinese New Year and the Asian population and number of Asian restaurants in NYC at census tract level. Twitter data was collected using Twitter API; the Asian population data were retrieved from U.S. Census Bureau; and the number of Asian restaurants was acquired by using Google Places API (Figure 14.5).

To collect Twitter data, first, a geographical bounding box of NYC is specified to collect geotagged tweets from 1 week before to 1 week after Chinese New Year. Second, we select a number of keywords (including "spring festival," "Chinese new year," "dumplings," "china lantern," and "red envelope") to filter these collected tweets, ultimately collecting 2453 tweets related to Spring Festival in New York City. Third, these filtered tweets are saved into a PostgreSQL table, containing detailed information such as "Tweet_ID," "UserName," "TimeCreated," "Lat," "Lon," "Hashtag," "Retweet," "ReplyTo," and "Text." Figure 14.6 gives an example of our Twitter data output. Note that, although the number of geotagged tweets is still very small

**FIGURE 14.5**
Asian population in New York City at census tract level (left) and the number of Asian restaurants in New York City at census tract level (right).

| | C | D | E | F | G | H | | I | J | |
|---|---|---|---|---|---|---|---|---|---|---|
| | Tweet_ID | UserScreen | TimeCreate | Lat | Lon | Hashtags | | Retweetnum | ReplytoTwe | Text |
| | 568525784521908224L | ninamarieox | 2015/2/19 | 40.5097 | -74.248 | | | 0 | None | We're going out for Chinese New Year!.. |
| | 568096495498608641L | ninamarieox | 2015/2/18 | 40.5098 | -74.248 | | | 0 | None | Heating up soup from a Chinese restaur: |
| | 568264026402238464L | ninamarieox | 2015/2/18 | 40.5098 | -74.248 | | | 0 | None | Thought it was 2014. then I said "no stup |

**FIGURE 14.6**
An example of Twitter data output.

in general (about 1% of all tweets), researchers using this information can be confident that they work with an almost complete sample of Twitter data when geographical boundary boxes are used for data collection. Figure 14.7 shows the geographical distribution of filtered collected Twitter data in New York City.

## 14.2.2 Density Analysis

In order to show areas where a higher density (or cluster) of activity occurs, a heat map is created by using a color gradient. A heat map is usually generated using kernel density estimation, which creates a density raster of an input point vector layer. Intuitively, the density is calculated based on the number of points in a location, with larger numbers of clustered points resulting in larger values. The most important step of this procedure is to determine the default search radius, or bandwidth, which consists of five steps (ESRI 2016a,b):

- Calculate the mean center of the input points.
- Calculate the distance from the (weighted) mean center for all points.

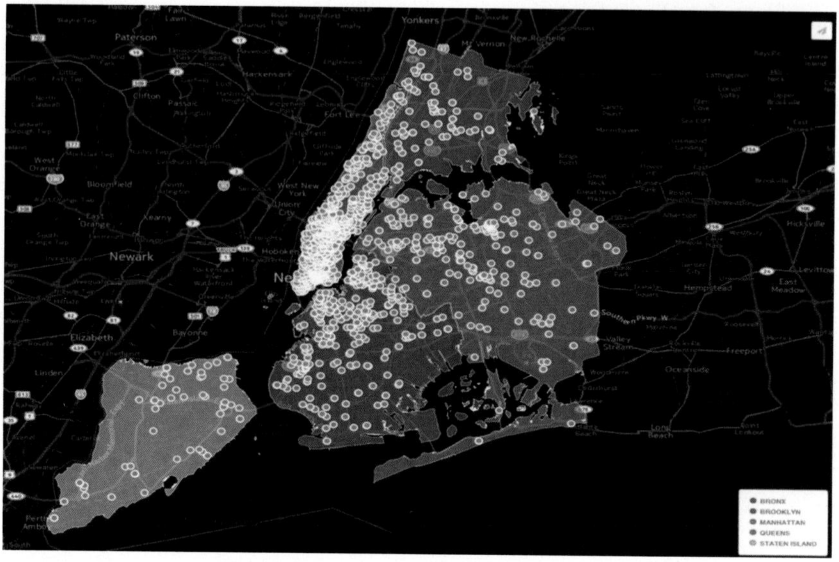

**FIGURE 14.7**
Geographical distribution of collected Twitter data in New York City. (From CartoDB.)

- Calculate the (weighted) median of these distances, $D_m$.
- Calculate the (weighted) standard distance, SD.
- Apply the following Equation 14.1 to calculate the bandwidth:

$$\text{Bandwidth} = 0.9 * \min\left(SD, \sqrt{\frac{1}{\ln(2)}} * D_m\right) * n^{-0.2} \qquad (14.1)$$

where SD is the standard distance, $D_m$ is the median distance, and $n$ is the number of points.

A heat map allows for easy identification of "hotspots" and clustered points. Figure 14.8 shows a heat map of filtered tweets in New York City created by CartoDB, a popular online GIS analysis application. As we can see, most of the tweets related to Spring Festival are concentrated in Manhattan and Queens.

## 14.2.3 Regression Analysis (OLS and GWR)

Regression analysis is used to model, examine, and explore spatial relationships and can help explain the factors behind observed spatial patterns. Ordinary least squares (OLS) is the best known of all regression techniques. It is also the proper starting point for all spatial regression analyses. It

**FIGURE 14.8**
Heat map of tweets in New York City (red represents high density, while blue represents low density).

provides a global model of the variable or process you are trying to under-stand or predict (early death/rainfall), creating a single regression equation to represent that process. Geographically weighted regression (GWR) is one of several spatial regression techniques, and increasingly being used in geography and other disciplines. GWR provides a local model of the vari-able or process to be understood/predicted by fitting a regression equation to every feature in the dataset. When used properly, these methods pro-vide powerful and reliable statistics for examining and estimating linear relationships.

In this case, OLS is applied as a starting point to explore the relationship between the number of involved Twitter users and demographic variables such as Asian population. To achieve this, data can be summarized by each census tract so that a regression analysis can be conducted at the census tract level. Two variables are selected to explain the number of tweets in each census tract: one is Asian population, and the other is the number of Asian restaurants, which is regarded as a proxy for Asians' real-world activities. A regular OLS model is performed initially, with the decision of whether or not to employ the GWR model depending on the OLS results.

**TABLE 14.1**

Result of OLS Analysis

| Variable | Coef | StdError | t_Stat | Prob |
|---|---|---|---|---|
| Number of restaurants | 2.906641 | 0.093339 | 31.140589 | 0 |
| Asian population | 0.001557 | 0.000218 | 7.151176 | 0 |

Table 14.1 shows the result of an OLS analysis. As can be seen, the $R^2$ value for this OLS model is only 0.35, which means only 35% of the variation in the dependent variable can be explained by these two independent variables. However, strong spatial autocorrelation is found in the spatial distribution of the standard residual for each census tract (Figure 14.9); thus, conducting a GWR analysis is deemed worthwhile. The global $R^2$ value for the GWR model is 0.699, which is much higher than that for the OLS model (Figure 14.10), implying that these relationships behave differently in different parts of the study area. Figure 14.10 shows the local $R^2$ value for GWR for each census tract. As can be seen, the highest local $R^2$ is clustered in Manhattan and part of Queens.

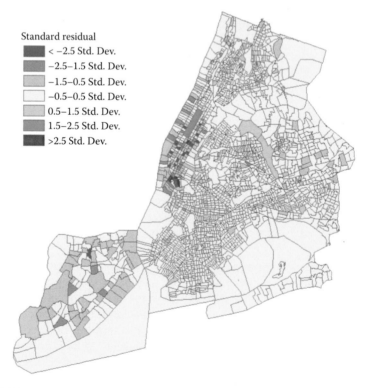

**FIGURE 14.9**
The standard residual of OLS of each census tract.

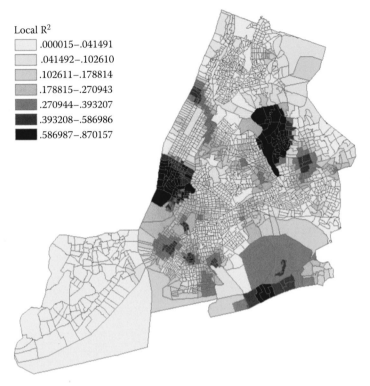

**Local R²**
| | |
|---|---|
| ☐ | .000015–.041491 |
| ☐ | .041492–.102610 |
| ☐ | .102611–.178814 |
| ☐ | .178815–.270943 |
| ☐ | .270944–.393207 |
| ☐ | .393208–.586986 |
| ■ | .586987–.870157 |

**FIGURE 14.10**
The local R² of GWR of each census tract.

## 14.3 Distributed GIS

Recently, with the development of data collocation technologies, data vendors, scientists, and even the general public scattered across the world are creating geospatial data on a daily basis. Traditional GIS running on a single machine can no longer adequately handle so many distributed resource and users. Accordingly, distributed GIS has been widely used to support distributed geospatial data management and processing. Climate science is one of the domains to which distributed GIS has been applied. Earth observation and model simulations are producing large amounts of climate datasets in a distributed computing environment. As a result, managing the complexity and magnitude of big climate data is not a feasible job, and sometimes exceeds the capability of climate scientists whose expertise is in data analysis. Furthermore, scientists rely on Internet-based collaborative climate research, such as model sharing and evaluation. Here, we use a Web-based geovisual analytical system as an example to show how distributed GIS can relieve climate researchers of the time-consuming tasks of data management and processing.

### 14.3.1 System Architecture

The three major tasks of the system are defined as managing, processing, and representing data; accordingly, the architectural components of the system will be (1) a data repository to store data or metadata from simulation, observation, and initial statistics, (2) an application Web server to provide data processing and high-level analytical functions, and (3) a Web-based client to perform simple analysis and display visualization results with interactive tools (Figure 14.11).

To efficiently manage this data, a spatiotemporal climate data repository is developed to uniformly manage the climate model output data by dynamically extracting its associated metadata into its own database. A scientific database is used to store the metadata for these data, such as content descriptors (variable names, resolution, extent, etc.) and storage information (format, file path in the file system, etc.). The data files themselves are stored as a data repository using a traditional file system. Other than raw simulated climate data, metadata and basic statistics (e.g., monthly/yearly mean) are generated and stored in the repository in order to speed up farther calculation. A standard Web service based on REST (Representational State Transfer) serves as a secure method for accessing the data repository though

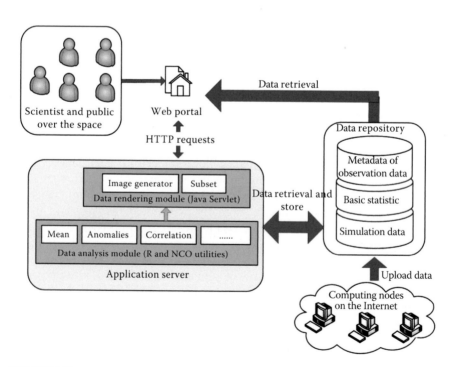

**FIGURE 14.11**
System architecture.

the Web, and using the standard REST ensures that data access remains interoperable.

Several analysis functions consulted by data scientists, such as the model validation Taylor diagram, are executed in the application server, which leverages high-performance computing resources. This allows large climate data to be processed much faster than a stand-alone analysis system. Data analysis requests are sent to the application server through HTTP request. The server side then executes analytical models and outputs the results as resulting figures, file paths, or values, which are returned to the client for rendering. On the client side, the system provides a user-friendly environment with geovisual analytical tools, which contain interactive tools, dynamic graphs/maps, and live-linked views of data representation.

This system has been implemented and is able to support several types of climate data, including the MERRA data introduced in Section 14.1, Climate Forecast System Reanalysis (CFSR) data, ECMWF Interim Reanalysis (ERA-Interim) data, CPC Merged Analysis of Precipitation (CMAP) data, Global Precipitation Climatology Project (GPCP) data, and ModelE simulation data, all of which are raster data. CFSR provides a global reanalysis of past weather from January 1979 through March 2011 at a horizontal resolution of 0.5°, and can effectively estimate the observed state of the atmosphere. ERA-INTRIM, which provides the reanalysis data from 1979 to the present, is an atmospheric model and assimilation system featuring improved low-frequency variability and stratospheric circulation analysis versus its previous generation, ERA-40. CMAP merges five kinds of satellite estimates (GPI, OPI, SSM/I scattering, SSM/I emission, and MSU) to provide the global gridded precipitation data from 1979 to near the present with a 2.5° spatial resolution. GPCP combines the data from rain gauge stations, satellites, and sounding observations to estimate monthly rainfall on a 2.5° global grid from 1979 to the present. ModelE is a general circulation model (GCM) developed by NASA GISS that simulates more than 300 variables on a global scale at a spatial resolution of 4° along parallels and 5° along meridians. The outputs are monthly binary data with a size of 16 MB.

## 14.3.2 User Interface

Figures 14.12 through 14.15 show the graphic interface of this WebGIS system with the following analysis and visualization algorithms/functions:

- Time series plotting (Figure 14.12): Users can select multiple data variables in multiple areas of interest (AOIs) for the same time period and plot the time series for better comparison.
- Correlation analyses for two variables with the same AOI (Figure 14.13): Display the relationships of any two variables or two AOIs using scatter plots.

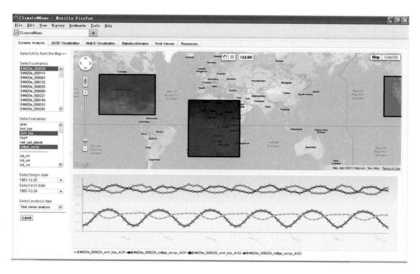

**FIGURE 14.12**
Time series plotting for two variables and two AOIs.

- Synchronous visualization of multiple parameters (Figure 14.14): When users interact with one map window, the other three map windows will simultaneously zoom to the same scale and geographic region to compare the selected variables.
- Data validation analysis (Figure 14.15): A Taylor diagram is a way of graphically summarizing how closely a pattern (or a set of patterns)

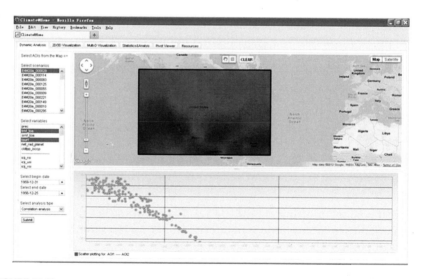

**FIGURE 14.13**
Correlation analyses for two variables.

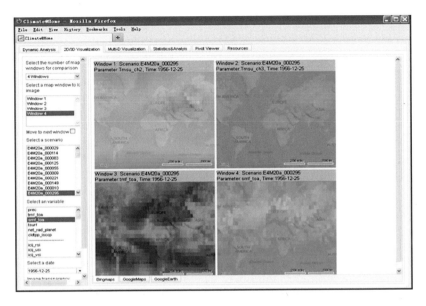

**FIGURE 14.14**
Four variables displayed in four windows.

matches observations. The position of each point appearing on the
plot quantifies how closely that model's simulated result pattern
matches observations. Users can input the spatiotemporal range
and select the climate data from different models for comparison.
The server end will then retrieve the specified data in parallel and

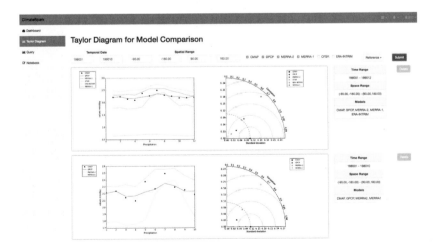

**FIGURE 14.15**
GUI for Taylor diagram service.

compute the parameters, such as standard deviation, root mean square, and correlation to incorporate in Taylor diagram. After these parameters are visualized in a Taylor diagram, the client will render the Taylor diagram images with the information about its spatiotemporal range and input datasets. The Taylor diagrams for different spatiotemporal ranges can also be displayed on the same page for intercomparison.

## 14.4 Spatiotemporal Thinking and Computing

The geographical world evolves in a four-dimensional fashion: three spatial dimensions and one temporal dimension. All phenomena on Earth are evolving in this four-dimensional world. Better understanding the evolution of integral space and time could help us understand our living environment and better prepare us to solve different problems and address emergency responses. This section uses an example of dust storm forecasting to illustrate how spatiotemporal patterns would help develop better thinking strategy and improve dust forecasting.

### 14.4.1 Problem: Dust Simulation and Computing Challenges

Dust storms are serious hazards to health, property, and the environment worldwide, especially in arid and semiarid regions such as northern China, the Middle East, the Saharan desert, and the southwestern United States (Knippertz and Stuut 2014). Dust storms rapidly reduce visibility, affect road and air traffic, increase power failures, and cause significant economic impact. In addition, dust storms allow pollutants and toxins such as salt, sulfur, heavy metals, and particulate matter to become resuspended in the atmosphere, resulting in negative long-term effects on human health (Wilkening et al. 2000).

To mitigate the hazardous impacts of dust storms, research is being conducted by international and national organizations to support comprehensive, coordinated, and sustained observations and modeling of dust storms. With the development of dust models, scientists have enhanced their understanding of dust event processes and have improved early warning capabilities for societal benefit (WMO 2011). In order to better support public health decision-making, the spatial and temporal resolution of dust storm simulations needs to be increased. This demand for higher resolution makes the model more computationally intensive. For example, if we run a coarse-resolution (1/3°) 72-hour dust model forecast for the U.S. Southwest using a single CPU, it will take about 4.5 hours to complete processing. When high-resolution simulation (e.g., 1/12°) output is needed, the computing time will increase by a factor of approximately 4 times in each of the three dimensions (latitude, longitude, and time steps), resulting in an overall increase of $4 \times 4 \times 4 = 64$ times, or approximately 12 days total to complete the processing—an unacceptable outcome. Therefore, one of the significant challenges for dust storm forecasting is to reduce computing time for a 1-day forecast to within an acceptable range (e.g., 2 hours) (Xie et al. 2010).

### 14.4.2 Methodology 1: Utilizing High-Performance Computing to Support Dust Simulation

In order to reduce the computing time of dust simulation to an acceptable range, the first method is to utilize high-performance computing to support dust simulation. High-performance computing can harness high-performance hardware such as multi-CPU-core computers or clusters to enhance computing capacity and reduce execution time for scientific models. Parallel processing, which is embedded in high-performance computing, partitions the model domain into subdomains so that each subdomain can be processed on a distinct CPU core, thus reducing the total processing time. During parallel processing, distributed-memory parallel computers need to communicate and synchronize so that the variable calculation is consistent for the entire model domain. For example, during the dust transition process, horizontal advection and lateral diffusion need to exchange dust concentration values among subdomains; therefore, communication among subdomains is critical for the dust simulation.

The dust model utilized here is a numerical weather prediction model (WRF-NMM), coupled with a dust module. The process of coupling dust modules is parallelized by decomposing the model domains on two levels (domain and subdomain) (Figure 14.16). For the dust simulation modules, MPI is used to communicate and exchange parameters and simulation results between patches. Communication is required between patches when a horizontal index is changed and the indexed value lies in the patch of a neighboring processor. Also, the boundary conditions need to be updated to maintain consistency after a periodic model running. Experiments have been conducted for a continuous 24-hour dust simulation to test the model running performance. Results show that it takes only about 9 minutes to run the model for a 6-hour simulation and about 36 minutes for a 24-hour simulation using 64 CPU cores.

### 14.4.3 Methodology 2: Utilizing Spatiotemporal Thinking to Optimize High-Performance Computing

The above example shows how parallelization improves the computing performance of dust simulation models. It decomposes the model domain into patches and tiles according to the default method for the parallelization platform (the middleware Message Passing Interface Chameleon,

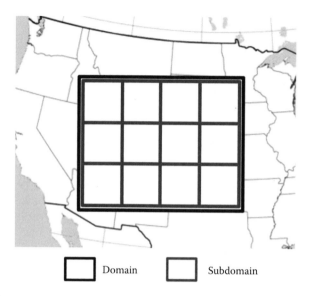

Domain     Subdomain

**FIGURE 14.16**
WRF-NMM domain decomposition for parallel running (NCAR, MMM Division, 2004).

version 2; MPICH2), which dispatches subdomains to the computing nodes. The following sections introduce further performance improvements by considering the spatiotemporal pattern of dust storm phenomena in terms of computing, especially the decomposition and scheduling approach.

### 14.4.3.1 Dust Storms' Clustered Characteristics: Scheduling Methods

The first spatiotemporal pattern of dust storm phenomena is that it is a clustered phenomenon with local dust concentration clustering. Therefore, dust storm simulation has to deal with clustered regions, and the scheduling method can take advantage of this characteristic. For example, if two computing nodes and eight subdomains are determined, then the first, third, fifth, and seventh subdomains will be dispatched to the first computing node and the remaining subdomains will be dispatched to the second computing node. MPICH2 uses the typical nonneighbor scheduling method. In the subdomain and computing nodes experiment, two computing nodes are utilized: half continuous subdomains are dispatched on the first computing node and the rest are dispatched on the other computing node in a neighbor scheduling fashion (Figure 14.17). In a cluster scheduling way, the number of communication pathways between different computing nodes will be less than the one using noncluster scheduling way.

In addition to the noncluster scheduling, Yang et al. (2011b) demonstrated that an HPC supporting geospatial sciences should leverage spatiotemporal principles and constraints to better optimize and utilize HPC

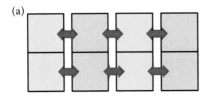 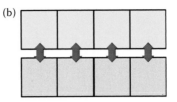

**FIGURE 14.17**
Two scheduling methods for dispatching eight subdomains to two computing nodes. (a) Non-cluster scheduling of 8 subdomains to 2 computing nodes (yellow and blue), (b) Non-cluster scheduling of 8 subdomains to 2 computing nodes (yellow and blue).

in a spatiotemporal fashion. Huang et al. (2013a) observe that performance improvement factors of approximately 20% on average could be achieved by using the neighbor scheduling method. This result suggests that it is better to dispatch neighbor subdomains to the same computing node to reduce the communication over computer networks.

### 14.4.3.2 Dust Storms' Space–Time Continuity: Decomposition Method

The second spatiotemporal pattern of dust storm phenomena is space–time continuity, meaning that it is generated, moves, evolves, and slows down in a continuous fashion. This space–time continuity results in the requirement that a numerical simulation exchange data among neighboring subdomains.

Yang et al. (2011b) also demonstrated that different decomposition methods result in different computing times. A notable reason is that dynamics are not consistent along the space, that is, velocities are relatively large near the poles and are much smaller in the North–South (meridional) direction than those in the East–West (zonal) direction (Nanjundiah 1998). Spatial principle 2, "spatial heterogeneity of physical phenomena," can be found in the noneven dynamical characteristic of atmospheric circulation. Therefore, communication needs differ among processors in the South–North (S–N) direction from those of the West–East (W–E) direction. In addition, different domain sizes along W–E and S–N directions result in different numbers of grid cells along these two directions. Thus, for the same degree of parallelization, different decompositions can result in different communication overheads.

Figure 14.18 shows the experiment results of various decompositions of 24 subdomains along S–N and W–E directions from the same domain. Results show that a one-dimensional decomposition in both longitude and latitude alone is a bad idea for parallel implementation (24 × 1 and 1 × 24), and that increased decomposition along longitude (S–N) direction is preferred, as 3 × 8 and 4 × 6 decompositions obtain higher performance than that of 8 × 3 and 6 × 4 decompositions (Yang et al. 2011b).

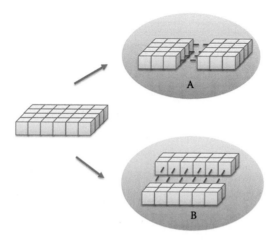

**FIGURE 14.18**
Different decomposition methods and their corresponding performance.

### 14.4.3.3 Dust Storm Events Are Isolated: Nested Model

The third spatiotemporal pattern of dust storm is that dust storm events are isolated or restricted in their spatiotemporal scope and can, therefore, be treated event by event. Various dust models have been developed to simulate the emission, transport, and deposition of dust storms using different resolution limitations and domain ranges. It would be ideal for a high-resolution model to simulate a large spatial domain, but this goal is difficult to accomplish due to computing capacity and memory consumption (Yang et al. 2011b).

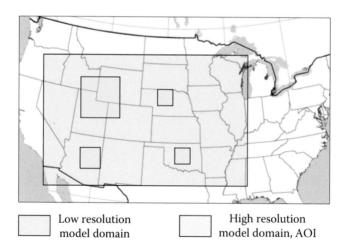

**FIGURE 14.19**
Low-resolution model domain area and subregions (AOI, area of interest,) identified for high-resolution model execution.

Therefore, Huang et al. (2013a) proposed an adaptive, loosely coupled strategy, which couples a high-resolution/small-scale dust model with a coarse-resolution/large-scale model. Specifically, the adaptive, loosely coupled strategy would (1) run the low-resolution model; (2) identify subdomains of high predicted dust concentrations (Figure 14.19); and (3) run the higher-resolution model for only those subdomains with much smaller area in parallel.

In this approach, high-resolution model results for specific subdomains of interest could be obtained more rapidly than an execution of a high-resolution model over the entire domain. Result shows the execution time required for different AOIs when HPC handles all AOIs in parallel, and it is expected to finish within 2.7 hours if all of the AOIs are simulated by the NMM dust model in parallel.

### 14.4.4 Methodology 3: Utilizing Cloud Computing to Support Dust Storm Forecasting

Dust storms have interannual variabilities and are typical disruptive events. The report from NOAA (2011) shows that the reported yearly frequency and percentage of the total time of dust storm in the United States varies each year. It is estimated that the total time of dust storms was generally less than 90 hours, representing less than 1% of a year, if it is assumed that each dust storm lasts an average of 2 hours. Therefore, a forecasting system for such events would require different computing and access requirements during different times of a year and even different hours within a day. Cloud computing is such a computing platform in that its computing capabilities can be rapidly and elastically provisioned, in some cases automatically, to scale up, and rapidly released to scale down. With the capability of providing a large, elastic, and virtualized pool of computational resources, cloud computing becomes a new and advantageous computing paradigm to resolve scientific problems traditionally requiring a large-scale, high-performance cluster (Huang et al. 2013b).

In Huang et al. (2013b), local cluster and cloud computing platforms (Amazon EC2) are tested and compared based on 15 computing tasks with different domain sizes for 24-hour forecasting. Result shows that Amazon cloud instances can complete most of these tasks in less time with the dedicated computing platforms for each task. The results indicate that cloud computing has great potential to resolve the concurrent intensity of the computing demanding applications.

## 14.5 Chapter Summary

This chapter introduces how GIS programming and algorithms are utilized to advance GIScience. Four examples were given to reflect the four fast-growing aspects of GIScience. The first one is developing new spatial data structure using climate data as an example, introducing a spatiotemporal index for speeding up climate data access. The second example introduced how big social media data can be analyzed to extract information of interest from the algorithm and modeling aspects. The third introduced how a distributed GIS can be developed to support online climate data discovery, navigation, and visual analytics. The fourth example discussed how to leverage spatiotemporal thinking and computing to improve dust storm simulation speed. These four examples demonstrate the advancements of GIScience using GIS programming and algorithms and are expected to serve as a bridge to advance future GIScience studies.

## PROBLEMS

*Requirement:*

- Analyze the problems solved in each of the four examples in this chapter to identify where programming can be used and how algorithms are developed for modeling objectives.
- Identify an academic GIScience journal paper, read and understand the paper, and try to lay out the aspects of the problem, data, algorithm, and solution.
- Discuss how to create a program to implement the solution analyzed in the journal paper.

*Deliverables:*

- Project Report
  - Explain the paper problem.
  - Explain the solution.
  - Explain how the algorithms learned in class can be adopted to develop the solution.
  - Keep it as simple but clear as possible, and try to use a diagram or picture in your report.

# References

Agarwal, D., Puri, S., He, X., and Prasad, S.K. 2012. A system for GIS polygonal overlay computation on Linux cluster-an experience and performance report. In *Parallel and Distributed Processing Symposium Workshops and PhD Forum (IPDPSW), 2012 IEEE 26th International*, 1433–9. IEEE, Shanghai, China.

Aho, A.V. and Ullman, J.D. 1972. *The Theory of Parsing, Translation, and Compiling*. Upper Saddle River, NJ: Prentice Hall, Inc.

Aji, A., Wang, F., Vo, H., Lee, R., Liu, Q., Zhang, X., and Saltz, J. 2013. Hadoop GIS: A high performance spatial data warehousing system over MapReduce. *Proceedings of the VLDB Endowment* 6(11):1009–20.

Arnold, K., Gosling, J., Holmes, D., and Holmes, D. 2000. *The Java Programming Language*, Vol. 2. Reading, MA: Addison-Wesley.

Benedetti, A., Baldasano, J.M., Basart, S. et al. 2014. Operational dust prediction. In *Mineral Dust: A Key Player in the Earth System*, eds. P. Knippertz, and W.J.-B. Stuut, 223–65. Dordrecht: Springer Netherlands.

Bondy, J.A. and Murty, U.S.R. 1976. *Graph Theory with Applications* (290). New York: Citeseer.

Bosch, A., Zisserman, A., and Munoz, X. 2007. Image classification using random forests and ferns. *International Conference on Computer Vision*, Rio de Janeiro, Brazil.

Bourke, P. 1988. *Calculating the Area and Centroid of a Polygon*. Swinburne University of Technology, Melbourne, Australia.

Chang, K.T. 2006. *Introduction to Geographic Information Systems*. Boston, MA: McGraw-Hill Higher Education.

Chen, Z. and Guevara, J.A. 1987. Systematic selection of very important points (VIP) from digital terrain models for construction triangular irregular networks. *Proceedings, AutoCarto 8*, ASPRS/ACSM, Falls Church, VA, 50–6.

Crooks, A., Croitoru, A., Stefanidis, A., and Radzikowski, J. 2013. Earthquake: Twitter as a distributed sensor system. *Transactions in GIS* 17:124–47.

Dale, M.R., Dixon, P., Fortin, M.J., Legendre, P., Myers, D.E., and Rosenberg, M.S. 2002. Conceptual and mathematical relationships among methods for spatial analysis. *Ecography* 25(5):558–77.

Dee, D. and National Center for Atmospheric Research Staff. eds. The Climate Data Guide: ERA-Interim. https://climatedataguide.ucar.edu/climate-data/era-interim (accessed June 9, 2016).

Dijkstra, E.W. 1959. A note on two problems in connexion with graphs. *Numerische Mathematik* 1:269–71.

Eckerdal, A., Thuné, M., and Berglund, A. 2005. What does it take to learn 'programming thinking'? In *Proceedings of the First International Workshop on Computing Education Research*, 135–42. ACM, Seattle, WA.

ESRI. 1998. ESRI Shapefile Technical Description. An ESRI White Paper, 34.

ESRI. 2016a. What Is ModelBuilder?, http://pro.arcgis.com/en/pro-app/help/analysis/geoprocessing/modelbuilder/what-is-modelbuilder-.htm (accessed September 9, 2016).

ESRI. 2016b. What Is ArcPy?, http://pro.arcgis.com/en/pro-app/arcpy/get-started/what-is-arcpy-.htm (accessed September 9, 2016).

Fowler, R.J. and Little, J.J. 1979. Automatic extraction of irregular network digital terrain models. *Computer Graphics* 13:199–207.

Fowler, M. 2004. *UML Distilled: A Brief Guide to the Standard Object Modeling Language.* Boston, MA: Addison-Wesley Professional.

Gittings, B.M., Sloan, T.M., Healey, R.G., Dowers, S., and Waugh, T.C. 1993. Meeting expectations: A review of GIS performance issues. In *Geographical Information Handling–Research and Applications*, ed. P.M. Mather, 33–45. Chichester: John Wiley & Sons.

Goodchild, M.F. 1992. Geographical information science. *International Journal of Geographical Information Systems* 6(1):31–45.

Gosselin, T.N., Georgiadis, G., and Digital Accelerator Corporation. 2000. Digital data compression with quad-tree coding of header file. U.S. Patent 6,094,453.

Grafarend, E. 1995. The optimal universal transverse Mercator projection. In *Geodetic Theory Today*, 51–51. Berlin, Heidelberg: Springer.

Guttman, A. 1984. R-trees: A dynamic index structure for spatial searching. *Proceedings of the 1984 ACM SIGMOD International Conference on Management of Data* 14(2):47–57. ACM.

Healey, R., Dowers, S., Gittings, B., and Mineter, M.J. eds. 1997. *Parallel Processing Algorithms for GIS*. Bristol, PA: CRC Press.

Hearnshaw, H.M. and Unwin, D.J. 1994. *Visualization in Geographical Information Systems*. Hoboken, NJ: John Wiley & Sons Ltd.

Hoeffding, W. 1963. Probability inequalities for sums of bounded random variables. *Journal of American Statistical Association* 58:13–30.

Huang, Q., Yang, C., Benedict, K., Chen, S., Rezgui, A., and Xie, J. 2013b. Utilize cloud computing to support dust storm forecasting. *International Journal of Digital Earth* 6(4):338–55.

Huang, Q., Yang, C., Benedict, K. et al. 2013a. Using adaptively coupled models and high-performance computing for enabling the computability of dust storm forecasting. *International Journal of Geographical Information Science* 27(4):765–84.

Hutchins, W.J. 1986. *Machine Translation: Past, Present, Future*, 66. Chichester: Ellis Horwood.

Hwang, K. and Faye, A. 1984. *Computer Architecture and Parallel Processing*, Columbus, OH: McGraw-Hill.

Jack, K. 2011. *Video Demystified: A Handbook for the Digital Engineer*. Burlington, MA: Elsevier.

Johnson, R.A. 1929. *Modern Geometry: An Elementary Treatise on the Geometry of the Triangle and the Circle*, 173–6, 249–50, and 268–9. Boston, MA: Houghton Mifflin.

Kanan, C. and Cottrell, G.W. 2012. Color-to-grayscale: Does the method matter in image recognition? *PloS One* 7(1):e29740.

Kernighan, B.W. and Ritchie, D.M. 2006. *The C Programming Language*. Upper Saddle River, NJ: Prentice Hall.

Khalid, M. 2016. Map, Filter and Reduce. http://book.pythontips.com/en/latest/map_filter.html (accessed September 3, 2016).

Knippertz, P. and Stuut, J.B.W. 2014. *Mineral Dust*. Dordrecht, Netherlands: Springer.

Lee, J. 1991. Comparison of existing methods for building triangular irregular network, models of terrain from grid digital elevation models. *International Journal of Geographical Information System* 5(3):267–85.

Li, Z., Hu, F., Schnase, J.L. et al. 2016. A spatiotemporal indexing approach for efficient processing of big array-based climate data with MapReduce. *International Journal of Geographical Information Science* 1–19.

Lien, D.A. 1981. *The Basic Handbook: Encyclopedia of the Basic Computer Language.* San Diego, CA: Compusoft Pub.

Linuxtopia. 2016. Set Operations. http://www.linuxtopia.org/online_books/ programming_books/python_programming/python_ch16s03.html (accessed September 3, 2016).

Longley, P.A., Goodchild, M.F., Maguire, D.J., and Rhind, D.W. 2001. *Geographic Information System and Science.* England: John Wiley & Sons, Ltd., 327–9.

McCoy, J., Johnston, K., and Environmental Systems Research Institute. 2001. *Using ArcGIS Spatial Analyst: GIS by ESRI.* Redlands, CA: Environmental Systems Research Institute.

Mitchell, J.C. 1996. *Foundations for Programming Languages* (1). Cambridge: MIT press.

Misra, P. and Enge, P. 2006. *Global Positioning System: Signals, Measurements and Performance Second Edition.* Lincoln, MA: Ganga-Jamuna Press.

Nanjundiah, R.S. 1998. Strategies for parallel implementation of a global spectral atmospheric general circulation model. *High Performance Computing, 1998. HIPC'98. 5th International Conference On* 452–8. IEEE.

Neteler, M. and Mitasova, H. 2013. *Open Source GIS: A GRASS GIS Approach* (689). New York: Springer Science and Business Media.

NOAA. 2011. Dust Storm Database. https://www.ncdc.noaa.gov/stormevents/ (accessed August 30, 2016).

Ostrom, E., Burger, J., Field, C.B., Norgaard, R.B., and Policansky, D. 1999. Revisiting the commons: Local lessons, global challenges. *Science* 284(5412):278–82.

Peng, Z.R. 1999. An assessment framework for the development of Internet GIS. *Environment and Planning B: Planning and Design* 26(1):117–32.

Pick, M. and Šimon, Z. 1985. Closed formulae for transformation of the Cartesian coordinate system into a system of geodetic coordinates. *Studia geophysica et geodaetica* 29(2):112–9.

Pountain, D. 1987. Run-length encoding. *Byte* 12(6):317–9.

Proulx, V.K. 2000. Programming patterns and design patterns in the introductory computer science course. *ACM SIGCSE Bulletin* 32(1):80–4. ACM.

Python. 2001a. Built-in Functions. https://docs.python.org/3/library/index.html (accessed September 3, 2016).

Python. 2001b. Errors and Exceptions. https://docs.python.org/2/tutorial/errors. html (accessed September 3, 2016).

Pythoncentral. 2011. Python's range() Function Explained. http://pythoncentral.io/ pythons-range-function-explained/ (accessed September 3, 2016).

PythonForBeginners. 2012. Reading and Writing Files in Python. http://www. pythonforbeginners.com/files/reading-and-writing-files-in-python (accessed September 3, 2016).

Raschka, S. 2014. A Beginner's Guide to Python's Namespaces, Scope Resolution, and the LEGB Rule. http://sebastianraschka.com/Articles/2014_python_scope_ and_namespaces.html (accessed September 3, 2016).

Rawen, M. 2016. *Programming: Learn the Fundamentals of Computer Programming Languages (Swift, C++, C#, Java, Coding, Python, Hacking, Programming Tutorials).* Seattle, WA: Amazon Digital Services LLC, 50.

Rew, R. and Davis G. 1990. NetCDF: An interface for scientific data access. *IEEE Computer Graphics and Applications* 10(4):76–82.

Ritter, N. and Ruth, M. 1997. The GeoTiff data interchange standard for raster geographic images. *International Journal of Remote Sensing* 18(7):1637–47.

Rodrigue, J. 2016. Network Data Models. Methods in Transport Geography. https://people.hofstra.edu/geotrans/eng/methods/ch2m3en.html (accessed June 22, 2016).

Rumbaugh, J., Blaha, M., Premerlani, W., Eddy, F., and Lorensen, W.E. 1991. *Object-Oriented Modeling and Design*, 199(1). Englewood Cliffs, NJ: Prentice Hall.

Samet, H. 1990. *The Design and Analysis of Spatial Data Structures*, 199. Reading, MA: Addison-Wesley.

Shepard, D. 1968. A two-dimensional interpolation function for irregularly-spaced data. In *Proceedings of the 1968 23rd ACM National Conference*, New York, 517–24. ACM.

Shimrat, M. 1962. Algorithm 112: Position of point relative to polygon. *Communications of the ACM* 5(8):434.

Stroustrup, B. 1995. *The C++ Programming Language*. Delhi, India: Pearson Education India.

Teampython. 2013. Conditioned Choropleth Maps (CCMaps) Generator. http://www.arcgis.com/home/item.html?id=a1c79e4cb3da4e0db9dc01b11fea9112 (accessed September 9, 2016).

Tu, S., Flanagin, M., Wu, Y. et al. 2004. Design strategies to improve performance of GIS web services. *ITCC* 2:444.

Tullsen, D.M., Eggers, S.J., and Levy, H.M. 1995. Simultaneous multithreading: Maximizing on-chip parallelism. *ACM SIGARCH Computer Architecture News* 23(2):392–403. ACM.

Van Rossum, G. 2007. Python programming language. In *USENIX Annual Technical Conference*, Santa Clara, CA, 41.

wiki.gis. 2011. Centroid. http://wiki.gis.com/wiki/index.php/Centroid (accessed September 3, 2016).

Wilkening, K.E., Barrie, L.A., and Engle, M. 2000. Trans-Pacific air pollution. *Science* 290(5489):65.

World Meteorological Organization (WMO). 2011. *WMO Sand and Dust Storm Warning Advisory and Assessment System (SDSWAS)—Science and Implementation Plan 2011–2015*. Geneva, Switzerland: WMO.

Xie, J., Yang, C., Zhou, B., and Huang, Q. 2010. High-performance computing for the simulation of dust storms. *Computers, Environment and Urban Systems* 34(4):278–90.

Yang, C., Goodchild, M., Huang, Q. et al. 2011a. Spatial cloud computing: How can the geospatial sciences use and help shape cloud computing? *International Journal of Digital Earth* 4(4):305–29.

Yang, C., Wong, D.W., Yang, R., Kafatos, M., and Li, Q. 2005. Performance-improving techniques in web-based GIS. *International Journal of Geographical Information Science* 19(3):319–42.

Yang, C., Wu, H., Huang, Q. et al. 2011c. *WebGIS Performance Issues and Solutions*, 121–38. New York, USA: Taylor & Francis.

Yang, C., Wu, H., Huang, Q., Li, Z., and Li, J. 2011b. Using spatial principles to optimize distributed computing for enabling the physical science discoveries. *Proceedings of the National Academy of Sciences* 108(14):5498–503.

Yang, C., Xu, Y., and Nebert, D. 2013. Redefining the possibility of digital Earth and geosciences with spatial cloud computing. *International Journal of Digital Earth* 6(4):297–312.

Zhang, J. 2010. Towards personal high-performance geospatial computing (HPC-G): Perspectives and a case study. In *Proceedings of the ACM SIGSPATIAL International Workshop on High Performance and Distributed Geographic Information Systems*, San Jose, CA, 3–10. ACM.

# Index